T0227869

EFFECT OF OPERATIONAL VARIABLES ON NITROGEN TRANSFORMATIONS IN DUCKWEED STABILIZATION PONDS

Effect of Operational Variables on Nitrogen Transformations in Duckweed Stabilization Ponds

DISSERTATION
Submitted in fulfilment of the requirements of
the Academic Board of Wageningen University and
the Academic Board of the UNESCO-IHE Institute for Water Education
for the Degree of DOCTOR
to be defended in public
on Wednesday, 23 March 2005 at 15:30 h in Delft, The Netherlands

by

JULIA ROSA CAICEDO BEJARANO
born in Cali, Colombia

CRC Press
Taylor & Francis Group
Boca Raton London New York

CRC Press is an imprint of the
Taylor & Francis Group, an **informa** business
A TAYLOR & FRANCIS BOOK

CRC Press
Taylor & Francis Group
6000 Broken Sound Parkway NW, Suite 300
Boca Raton, FL 33487-2742

First issued in hardback 2017

© 2005 by Taylor & Francis Group, LLC
CRC Press is an imprint of Taylor & Francis Group, an Informa business

No claim to original U.S. Government works

ISBN 13: 978-1-138-41887-5 (hbk)
ISBN 13: 978-0-415-37554-2 (pbk)

Visit the Taylor & Francis Web site at
http://www.taylorandfrancis.com

and the CRC Press Web site at
http://www.crcpress.com

Dedication

I dedicate this thesis to:

Jesus, who gives me his light,
Jorge, my beloved father, who is always with me from the eternity,
Servia, my mother, who has given me her love,
Maria del Rosario, Luis Enrique, Claudia Alejandra, and Juan
Carlos, who have given me their support and love.

God bless them all

Dedicatoria

Esta tesis la dedico a:

Jesús, quien siempre me ilumina,
Jorge, mi amado padre, quien siempre me acompaña desde la
eternidad,
Servia, mi madre, quien me ha entregado su amor,
Maria del Rosario, Luis Enrique, Claudia Alejandra y Juan Carlos,
quienes me han dado amor y apoyo.

Dios los bendiga a todos

Acknowledgements

To GOD, who has supported me during this journey in every step of the way

To the Dutch Government (SAIL) for financially supporting this research within the scope of the collaboration project between UNESCO-IHE Institute for Water Education (The Netherlands) and Universidad del Valle – (Colombia).

To my promoter Professor Huub Gijzen, I am indeed very grateful for his guidance, kindness and friendship during the research work and write-up of the thesis, for his continuous encouragement and support, for his useful discussions, comments and ideas in Delft and in Cali.

To my co-promotor Dr. Peter van der Steen, I am also in debt for his constructive discussions and advice, and for the long hours spent in reviewing the manuscripts.

To Dr Maarten Siebel, I thank for his useful discussions during the writing of the proposal and initiation of the experimental work.

To the Universidad del Valle for giving me the opportunity to undertake this study. Many thanks to Alberto Galvis for his collaboration as the Colombian coordinator of the project between UNESCO-IHE Institute for Water Education (The Netherlands) and Universidad del Valle –(Colombia). Many thanks to Hernán Materón, Director of the School of Natural Resources and Environment- Engineering Faculty- Universidad del Valle, for his support to finish this work.

To the dedicated staff and students from both the Universidad del Valle and UNESCO-IHE. Thanks to Diego Mayor, Janeth Sanabria, Juan Pablo Silva, Isabel Cristina Yoshioka, Carlos Espinosa, Clara Glas, Mario Márquez, Diego Paredes, Amarilis Herrera, Jihad Sasa, Jenny Soto, Angy Quintero, Deisy Timaná, Alba Pérez, Luz Marina Londoño, Aida Lima, Zandra Palacios, Darío Agudelo, Luis Saavedra, Edward Murillo, Fred Kruis. Many thanks also to my colleagues Omar Zimmo and Esi Awuah.

To ACUAVALLE's staff, I thank to Ir Alex Sánchez, Farid Montenegro, and Diego Corrales. I want to give special thanks to the laboratory staff in the Ginebra's Research Station, Yimmer Vélez, Haider Pérez, Evelio Castellanos, John Freddy Trujillo, for their valuable contribution to the experimental work.

To my family, who has given me a lot of love throughout the years. And last, but not least, to my friend Leoni Zweekhorst and her family, for all their love and support during my stay in Holland.

To all those whom I might have failed to mention on this page, once again my deepest thanks and appreciation.

List of abbreviations

NH_3	Ammonia
NH_4^+	Ammonium
NH_4^+-N	Ammonium nitrogen
et al.	And others
av	Average
BOD	Biochemical oxygen demand
BOD_5	Biochemical oxygen demand (5 day)
cal	Calories
cm	Centimeter
COD	Chemical oxygen demand
d	Days
DO	Dissolved oxygen
DSP	Duckweed stabilization pond system
EU	Effluent of UASB reactor
FC	Faecal coliforms
K_B	First order faecal coliform degradation constant
k_{BOD}	First order biochemical oxygen demand degradation constant
Q	Flow rate
i.e.	For example
FW	Fresh weight
g	Grams
GNP	Gross national product
ha	Hectares
HRT	Hydraulic retention time
Kg	Kilograms
m	Meter
mg l^{-1}	Milligrams per liter
mg/l	Milligrams per liter
NO_3	Nitrate
NO_3-N	Nitrate nitrogen
NO_2	Nitrite
NO_2-N	Nitrite nitrogen
NOx-N	Nitrite nitrogen + Nitrate nitrogen
N	Nitrogen
N_2	Nitrogen gas
n	Number of samples
Org-N	Organic nitrogen
P	Phosphorus
RW	Raw wastewater
RGR	Relative growth rate
s	Second
cm^2	Square centimeter
m^2	Square meter
SD	Standard deviation
se	Standard error
T	Temperature
t	Time

TKN	Total Kjeldahl nitrogen
TN	Total nitrogen
TP	Total phosphorus
TSS	Total suspended solids
UASB	Up-flow anaerobic sludge blanket
WSP	Waste stabilization pond
WWT	Wastewater treatment

Greek

μS	Micro-Siemens
μE	Micro-Einsteins

Table of content

Chapter 1

Introduction

Chapter 1

Introduction

Rapid population growth, industrial development and intensified agricultural production have generated increasing amounts of wastewater, which are causing contamination of receiving water bodies in many places around the world at an alarming rate. The current concept of water supply is to distribute large amounts of water to the communities for indiscriminate use with the subsequent need to collect and to transport the produced wastewater out of the urban areas to be treated in centralised wastewater treatment plants. A new concept has to be developed to control the growing water pollution problems around the world (Graaf *et al.*, 1997; Larsen and Gujer, 1997; Gijzen and Ikramullah, 1999). This will be possible only with a combination of different approaches: firstly, via the development and implementation of strategies to reduce or to prevent wastewater generation. Secondly, via the development of technologies that, effectively remove pollutants, facilitate reuse and, at the same time, are technically and economically accessible by most countries. These changes are not expected in a short term as an enormous amount of money will have to be spent in order to modify the existing infrastructure in water supply and sewerage systems. In the meantime strategies for water saving and pollution prevention should be implemented, and sustainable treatment technologies for wastewater treatment should be developed.

There is a diversity of conventional technologies available for removal of pollutants from the wastewater. Most of these technologies are aerobic alternatives with high building cost, high energy consumption and requirements of skilled personal for operation and maintenance. As a consequence, only countries with a high gross national product (GNP) can afford these options. The urgent need for food production has generated an intensive agricultural activity which requires huge amounts of fertilizers obtained via industrial fixation of atmospheric nitrogen. The industrial fixed nitrogen represents 37% of the natural fixed nitrogen which causing a dramatic imbalance in the global nitrogen cycle and the increasing levels of eutrophication in water bodies worldwide (Gijzen, 2001). High investments in wastewater treatment plants during the last decades have greatly reduced the organic loading of receiving water bodies in different countries. However, many of the existing wastewater treatments plants are not equipped to remove the nutrients like nitrogen. Lately stricter regulations for nitrogen removal have been introduced, especially for discharge of into water courses (EU criteria: < 10 mg N l^{-1}). For effluent reuse in irrigation a reduction of nitrogen concentration to 15-25 mg N l^{-1}, depending on crop type, will be sufficient. Traditionally nitrogen has been removed from wastewater through processes like nitrification-denitrification, ammonia volatilization and microbial biomass uptake. The use of aquatic macrophytes in wastewater treatment is offering an attractive alternative to remove nitrogen and to recover it in the form of valuable vegetable biomass to be reuse as a source of protein.

Wastewater management in low-income countries

The wastewater treatment and management in countries with a low GNP is even worse than in the developed world. The unequal expansion of water supply coverage compared to the expansion in sanitation services generates the contamination of surface and ground waters with the subsequent environmental deterioration. Latin-America has a population of 511 million inhabitants (75.3 % urban population, 24.7 % rural population). A study performed by CEPIS-OPS-OMS (2004) for 60% (293 millions) of the total population on treatment systems and re-use of wastewater showed that 69% has sewer systems, while 27% has treatment systems. In Colombia, with 41 million inhabitants, the figures are very similar: 72% urban population, 28% rural population, 86% with sewer system, 31% with wastewater treatment system. The environmental legislation affects the industrial sector since some years only, but at the municipal level the situation is different. At the moment the larger cities are pushing strongly their pollution control programmes and are searching for feasible alternatives to solve the wastewater problem. Many of the middle sized and small cities do not have any treatment. In many other Latin-American countries, even though regulations do exist, these are not enforced. One of the main reasons for this is the limited economic resources to cover the high cost of sewerage and conventional wastewater treatment technologies. Of the 293 million people evaluated by CEPIS-OPS-OMS (2004) 32 % are living in big cities (> 1 million inhabitants), 25% in medium sized cities (1.000.000 – 100.000 inhabitants) and 41.5 % in small cities (100.000 – 2000 inhabitants). The general trend is to use conventional systems for big cities, but for medium and small sized cities non-conventional systems are often considered. Therefore, there is an urgent need to develop and improve low cost technologies for wastewater treatment that are within the economic and technological capabilities of a developing country like Colombia. At the same time these technologies should be reliable and effective in removing a wide range of pollutants. Furthermore, it would be ideal if these technologies can provide resource recovery like the generation of biogas (energy production), high quality biomass (animal fodder), or effluent fit for irrigation. At the moment, no technological packages appear to be readily available

World Water Vision has defined the target to achieve full coverage of water supply and sanitation by year 2025 (Cosgrove and Rijsberman, 2000). In Latin-America where the percentage of people connected to sewerage is high, wastewater treatment will be an important factor to reach the Millennium Development Goals in this part of the world. Therefore, the strategies for wastewater management have to take advantage of more effective, low tech and low cost technologies. Gijzen (2003) proposed a 3-step strategic approach for urban water management. First: pollution prevention or waste minimization through the reduction of water and chemicals going into the wastewater. Second: treatment for reuse in order to recover the useful materials always present in the wastewater. Third: planning of effluent disposition in order to stimulate the self purification capacity of the receiving water bodies. This is a holistic approach should lead to the development of more sustainable technologies for wastewater management.

Nutrient recovery through the use of macrophytes for wastewater treatment.
Scientists and engineers from several countries have paid attention to the potential of aquatic macrophytes to treat and recycle pollutants from municipal and industrial wastewater (Araujo, 1987; Brix and Schierup, 1989; Rao, 1986). These plants have the capacity to assimilate nutrients and to convert these directly into valuable biomass (Reed *et al.*, 1995). Various studies have reported the use of water hyacinth (*Eichhornia crassipes*), pennywort (*Hydrocotyle umbelata*), water lettuce (*Pistia stratiotes*) and duckweed (*Lemnaceae*) for the efficient removal of nutrients. Water hyacinth has been used most widely due to its high nutrient uptake capability (Reddy and Smith, 1987), but no economically attractive application of the generated plant biomass has been identified so far. Besides, water hyacinth only grows efficiently in tropical climates, thereby restricting its use in temperate climates.

Different authors have proposed the use of duckweed-based systems for the efficient and low cost treatment of domestic wastewater at urban or rural levels (Zirscky and Reed, 1988; Skillicorn *et al.,* 1993; Oron, 1994; Gijzen, 2001). These systems are stabilization ponds where the water surface is completely covered by a duckweed mat. The main treatment processes, which occur in the system, are sedimentation, aerobic and anaerobic bacterial degradation, and nutrient uptake by the plants.

Experience has shown that no single technology can offer an optimum treatment for the different components to be removed in wastewater like organic matter, suspended solids, nutrients and pathogens and to recover valuable resources like nitrogen. Therefore an adequate combination of different technologies in an integrated system could convert a wastewater treatment into an attractive sustainable system (Fig. 1). Anaerobic treatment in the first stages of the system will reduce considerably the organic matter in the wastewater and convert it into methane, which can be used as a fuel. The effluents of anaerobic treatment should be post-treated to meet discharge standards. Post-treatment of anaerobic effluents could be performed in macrophyte ponds for nutrient recovery in the form of high quality biomass that can be used for aquaculture and animal feed. The treated effluent can be used in irrigation. A system that generates these by-products increases the feasibility and sustainability of pollution control programs. Furthermore, the products may help to address the increasing need for food production in the world.

Researchers from UNESCO-IHE have been doing extensive research on duckweed ponds, jointly with other institutions around the world, including Bangladesh (Gijzen and Ikramullah, 1999), Yemen (Al-Nozaily, 2001), Colombia (Caicedo *et al.*, 2002), Palestine (Zimmo *et al.*, 2003), Zimbabwe (Nhapi, 2004) and Egypt (El-Shaffai, 2004). The work has been concentrated on the development of knowledge on the processes occurring within the ponds, with respect to organic matter, nitrogen, phosphorus and pathogen removal. Also the valorisation of recovered biomass in aquaculture was studied (El-Shafai *et al.*, 2004). In the process of development of duckweed pond technology, further research is needed to study the effect of different operational variables, like anaerobic pre-treatment, introduction of aerobic zones and pond depth on the performance of the system in terms of

4

treatment efficiencies, nutrient recovery and biomass production. This will be the challenge of the present research.

Duckweed characteristics.
Before discussing duckweed ponds as an option for wastewater treatment, some information about the characteristics of the plant is presented. Duckweed is scientifically known as the family of *Lemnaceae*, a family of aquatic floating plants, which consists of 4 genera: *Lemna, Spirodela, Wolffia* and *Wolffiella* (Fig. 2). It is a flowering plant with a very simple structure, with a fusion of leaves and stems called 'fronds' ranging in size between 0.1 cm. and 1.5 cm. Each frond has two meristemic

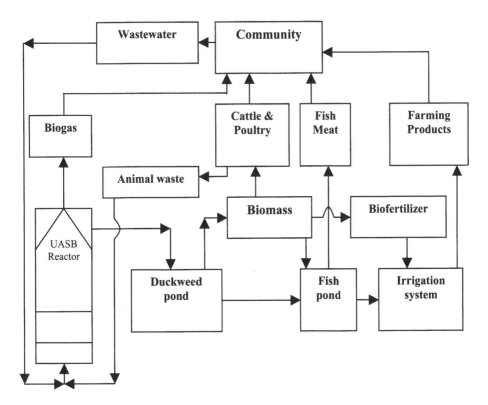

Fig. 1. Integrated system consisting of a UASB reactor, duckweed ponds, fish ponds and crop irrigation using wastewater treatment and resource recovery.

regions that alternatively produce new fronds. The relative high growth rates reflect on a good potential for nutrient uptake. The uptake of nutrients from the water is very efficient because the entire plant is used for this purpose, whereas other higher plants only use the root system. The absorption of nutrients and water is done mainly through the lower epidermis of the fronds (Ice and Couch, 1987; Landolt and Kandeler, 1987).

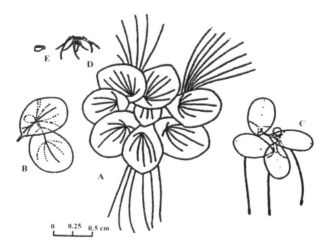

Fig. 2. Duckweed morphology and genera: A: *Spirodela*, B and C: *Lemna*, D: *Wolffiella*, E: *Wolfia*.

The protein content of *Lemnaceae* has been reported to be one of the highest within the plant kingdom (Table 1). Under ideal conditions a protein content of more than 40% can be achieved (Landolt, 1986; Skillicorn *et al.*, 1993). The high content of different amino acids, enzymes and vitamins and low content of fibres make duckweed a good candidate as an animal feed (Culley and Epps, 1973; Harvey and Fox, 1973; Mbagwu and Adeniji, 1988). Duckweed can also act as a good organic fertiliser because the level of nitrogen in the plant. Harvested duckweed, if grown on domestic wastewater free from heavy metals and other hazardous compounds, can be used as an agricultural fertiliser and in the production of high quality compost. An alternative use could be the generation of biogas from anaerobically digested duckweed.

Table 1. Variations of various components in *Lemnaceae* (Landolt and Kandeler, 1987)

Compound	% of dry weight
Proteins	6.8 - 45.0
Lipids	1.8 - 9.2
Crude fibres	5.7 - 16.2
Carbohydrates	14.1 - 43.6
Ash	12.0 - 27.6

Duckweed is known to grow under different environmental conditions and to be very resistant to a wide range of temperature, pH, nutrient concentrations as well as droughts, pests and diseases (Landolt and Kandeler, 1987). In comparison with other aquatic macrophytes duckweed is more tolerant to low temperatures than water hyacinth and easier to harvest than algae or water hyacinth. *Lemnaceae* are found in sunny and shady conditions. *Lemnaceae* has been found in waters containing high content of organic matter (Landolt, 1986). Its effect on growth is not well known.

Organic matter may play a role as a pH buffer, as a chelating agent or as a supplier of amino acids and vitamins for heterotrophic growth. Uptake of amino acids and other organic compounds was observed by Datko and Mudd (1985, as cited by Landolt, 1986). Furthermore, duckweed has been reported to grow in complete darkness if organic substances such as sugar are present in the medium (Landolt, 1986).

In relation to quality of water, the family of *Lemnaceae* is found in many different fresh and brackish waters and is able to grow in a wide range of nutrient concentrations (Table 2). In general, *Lemnaceae* do not grow on oligotrophic waters, they have high nutrient requirements and are resistant to relatively high salinity (Pott, 1980; Starfinger, 1983, as cited by Landolt, 1986; Oron *et al.*, 1985). This resistance of duckweed to high salinity could be an important factor in the application of duckweed-based systems in the reduction of conductivity to make water suitable for irrigation.

Table 2. Range of nutrient content in natural waters where *Lemnaceae* prevail (in mg l^{-1}, conductivity in μS cm^{-1}, Landolt, 1986).

Characteristics of water content	Absolute range	Range of 95% of the samples
pH	3.5 - 10.4	5.0 - 9.5
Conductivity	10 - 10900	50 – 2000
Ca	0.1 - 365	1.0 – 80
Mg	0.1 - 230	0.5 – 50
Na	1.3 - >1000	2.5 – 300
K	0.5 - 100	1.0 – 30
N	0.003 - 43	0.02 – 10
P	0.000 - 56	0.003 – 2
HCO$_3^-$	8 - 500	10 – 200
Cl	0.1 - 4650	1 – 2000
S	0.03 - 350	1 – 200

The pH may influence the availability of essential minerals for the plants like P, Fe, Mo, Zn and Mn, or the solubility of toxic materials (Buckman and Brady, 1969, as cited by McLay, 1976). It may also affect the behaviour of sensitive root cells due to the toxicity of the hydrogen ion itself (Ullrich *et al.*, 1984; Ingermarson *et al.*, 1987). McLay (1976) studied the pH effect on the growth rate of three species of duckweed *Spirodela oligorrhiza, Lemna minor, and Wolffia arrhiza*. The estimated lower, optimum and upper limits for each species were: *Spirodela* 3-7-10, *Lemna* 4-6.2-10 and *Wolffia* 4-5-10. The pH will determine the predominant form of N and S in the chemical equilibrium of NH_4^+-NH_3 and H_2S-S^-, which are common products of the anaerobic process. The un-dissociated form of these compounds (i.e. NH_3 and H_2S) is known to be toxic to plants. As the anaerobic process is included as one of the main variables in this project, special attention will be given to the relation pH-ammonium nitrogen and its effects on the duckweed.

The biology of duckweed has been extensively studied (Landolt, 1986; Landolt and Kandeler, 1987; Gijzen and Khondker, 1997). It has been very useful for a variety of bioassay studies at laboratory scale (Clark *et al.*, 1981; Wang, 1986; 1990; Smith and Kwan, 1989). Despite high availability of information on the biology of duckweeds, there is a need for more studies with respect to the use of duckweed in wastewater treatment and the possible reuse of the plant biomass as an animal feed or in other applications.

Duckweed ponds for wastewater treatment.

Although the use of duckweed ponds (Fig. 3) in wastewater treatment is rather recent, a number of authors concluded that in general, duckweed ponds are a feasible treatment method (Selvan *et al.*, 1992; Alaerts *et al.*, 1996; Gijzen and Ikramullah, 1999; Al-Nozaily, 2001; Zimmo, 2003; El-Shafai, 2004). As mentioned in the previous section, duckweed presents some attractive characteristics for this purpose, it grows very fast and as a consequence accumulates nutrients rapidly, it can be harvested quite easily, thus lowering harvesting cost and it possesses a good economic resource potential (Skillicorn *et al.*, 1993; Gijzen and Khondker, 1997).

Duckweed growth is inhibited at extreme high plant densities, which reduces photosynthesis and consequently affects the yield. Therefore, frequent harvesting is necessary. However, light can penetrate the water column if the density becomes too

Fig. 3. Photograph of an experimental duckweed stabilization pond in Ginebra-Colombia.

low and algae will grow and compete with duckweed for nutrients. Algae may release inhibitory substances for the duckweed into the water, and some species grow around the roots and over the fronds, causing decay and finally the death of the fronds. Consequently the density should be maintained at optimum levels. Skillicorn *et al.* (1993) recommended a range of duckweed biomass density between 400 and 800 g of fresh weight per m^2. Due to the importance to maintain a full cover over the water surface, the effect of wind on the small plants is one of the main constraints of the duckweed technology. To reduce the effect of the wind on the mat, different pond designs have been proposed, for example to install wooden grits on the surface of the lagoons or to build the ponds as narrow long channels in a zig zag structure.

Duckweed has been used to treat different types of waters, *i.e.* effluent of algae ponds, aquaculture system effluents, secondary municipal effluents, stabilisation ponds, domestic wastewater, dairy wastewater, etc. Duckweed can also be useful in the removal of nutrients from different animal manures. Porath and Pollock (1982) evaluated the potential of *Lemna gibba*, as a biological ammonia stripper. Their results indicated that the process of uptake was very active with preference for ammonia over nitrate. In an axenic culture, *Lemna gibba* absorbed 50% of the ammonia present at a level of 2 mg l^{-1} in 5 hours, while the nitrate level of 620 mg l^{-1} remained constant. The circulation of fishpond effluent with 10 mg l^{-1} NH_4^+-N under a duckweed mat promoted an uptake of 90% of the ammonia present within 48 hours. The direct conversion of ammonia by duckweed into plant protein, rather than protein production based on nitrate reduction, is considered an abridgement of the nitrogen cycle. Kawabata and Tatsukawa (1986) evaluated growth of duckweed and nutrient removal in a rice paddy field irrigated with secondary treated sewage effluent. It was concluded that duckweed growth plays a beneficial role as a mitigating agent for excessive nutrient supply to rice plants as well as a purifier of the sewage effluent.

Edwards *et al.* (1990) investigated the direct use of wastewater to feed a Tilapia fishpond, and compared this to feeding a fishpond with wastewater that had passed through a duckweed pond. The second system was more efficient in terms of fish production, as fresh fish weight was double compared to the first system, although the total area requirements were higher. In addition, the indirect use of wastewater is safer from a public health point of view. Selvam *et al.* (1992) did an extensive research in Bangladesh with fresh water fishponds. They obtained an average production of 100 kg of fish per ton of fresh duckweed and a fish yield of 10 ton ha⁻¹yr⁻¹. This yield is 50 to 70% higher than the fish yield in local fish ponds to which only fertiliser was applied to promote growth of phytoplankton for fish feed.

Oron *et al.* (1985) found that the removal efficiency of the major pollutants of domestic wastewater reached 50 to 60%, working with outdoor experiments conducted in mini-ponds. Alaerts *et al.* (1996) obtained removal efficiencies of 90-97 % for COD, 95-99 % for BOD_5, 74 for TKN and 77 % for TP, in a duckweed covered lagoon treating the wastewater of 2500 inhabitants. Steen *et al.* (1999) found poor pathogen removal in duckweed ponds treating the effluent of a UASB reactor, but they obtained improved efficiencies when combining duckweed ponds and algae ponds. With the combined system they found 56% of nitrogen removal with an overall retention time of 4.2 d (Steen *et al.*, 1998). Zimmo *et al.* (2002) studied the performance of a series of four duckweed ponds (28 days of HRT) obtaining annual average removal efficiencies of 92% BOD, 71% TSS, 54% N, 74-61% P. El-Shafai (2004) evaluated the performance of a series of three duckweed ponds (HRT = 15 days) treating effluent of a UASB reactor. The removal efficiencies during the warm season were: 93% COD, 96% BOD_5, 91% TSS, 85% TKN, 78% P.

The use of duckweed ponds in wastewater treatment has the following advantages:
- Low operational costs.
- Low energy requirements.
- Resistance to shock loading.
- Duckweed is easy to remove/harvest.
- Only few pests affect duckweed.
- Duckweed reduces mosquito growth, because the water surface is completely covered by the plants.
- Duckweed controls algae growth, because the cover on the surface of the water does not allow light to pass through. Therefore lower total suspended solids are expected in the final effluents (Steen, 1999).
- Duckweed controls odour release from ponds (Steen *et al.*, 2003)
- Duckweed reduces evaporation as compared to free surface water bodies (Oron *et al.*, 1985).
- Building and operational cost of the ponds are lower than for other types of wastewater treatment processes.
- There is good potential for resource recovery by harvesting and utilising the biomass produced as an animal fodder.

The main disadvantages of duckweed ponds are:
- The relatively high area requirement.
- Growth reductions at temperatures lower than 15°C.
- Less efficient removal of bacterial pathogens than in algal ponds (Steen *et al.*, 1999: Zimmo *et al.*, 2002).

Duckweed technology appears to present a good alternative for wastewater treatment in tropical countries like Colombia where temperature will not be a problem in most places. Area requirements have to be reduced to increase the feasibility of applying this technology, especially in urban areas of medium size and large size. Combination of duckweed ponds with high-rate pre-treatment systems may be a solution to overcome this limitation.

Anaerobic pre-treatment of wastewater.
During anaerobic treatment organic material is converted biologically by bacteria, under anaerobic conditions (absence of oxygen), to methane (CH_4) and carbon dioxide (CO_2).

A consortium of micro-organisms carries out the biological conversion of the organic matter (Fig. 4). One group of micro-organisms is responsible for hydrolysing organic polymers and lipids to basic structural building blocks, such as monosaccharides, amino acids, and related compounds. A second group of bacteria ferments the breakdown products to simple organic acids; the most common is acetic acid. These organisms are facultative and obligate anaerobic bacteria and are often identified as acidogens or acid formers. A third group converts the acids into acetate and the fourth group of micro-organisms converts the hydrogen and acetate into methane gas and carbon dioxide. The bacteria responsible for this conversion are strict anaerobes and are called methanogenic bacteria (Metcalf and Eddy, 1991). To maintain a stable anaerobic process the system should be void of dissolved oxygen and free from inhibitory concentrations of constituents such as heavy metals and sulphides. Sufficient alkalinity should be present and pH should range from 6.6 to 7.6. Nutrients present in the organic compounds will be released during hydrolysis and nitrogen compounds will be reduced to ammonium nitrogen.

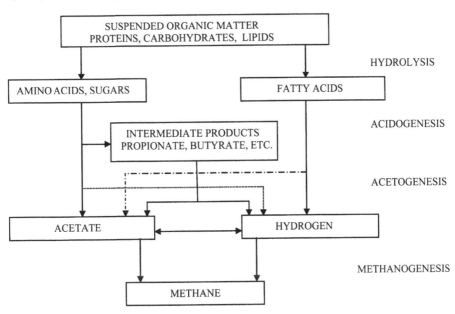

Fig. 4. Reaction sequence for the anaerobic processes (Adapted from Handel and Lettinga, 1994).

The disadvantages and advantages of anaerobic treatment of organic waste, as compared to aerobic treatment (with the presence of molecular oxygen), stem directly from the low growth rate of the methanogenic bacteria. Low growth rates require a relatively long microbial detention time in the reactor for adequate waste stabilisation to occur. However, the low growth yield signifies that only a small portion of the degradable organic waste is being synthesised into new cells. As a result a low production of reasonably well stabilised sludge is obtained. This results into reduced sludge management costs for anaerobic systems.

In the past fifteen years a number of 'modern' or high rate anaerobic reactors have been developed to overcome the limitation of the long retention times needed by the low rate anaerobic reactors. Modern systems can be characterised by the fact that they have a mechanism for sludge retention (Haandel and Lettinga, 1994). One example of the last case is the Up-flow Anaerobic Sludge Bed (UASB) reactor. In this reactor the wastewater is introduced in the bottom and it flows upwards through a sludge blanket composed of biologically formed granules or particles. Biodegradation processes occur as the wastewater comes in contact with the granules. Experience gained with UASB systems so far is that the advantages are considerable: low construction and operational costs compared to conventional systems, relatively small size, simplicity of construction and operation and production of little and well stabilised sludge (Haandel and Catunda, 1997).

An important drawback of the UASB reactor is that the effluent often does not comply with the environmental and legal requirements. The UASB reactor can be regarded as an advanced primary treatment with good organic matter removal but with negligible effect on nutrients or pathogens. In most cases it is necessary to apply post-treatment in order to upgrade the effluent quality before final discharge or re-use. The combination of anaerobic pre-treatment followed by macrophyte-covered stabilization ponds has been proposed by Gijzen (2001, 2002) for effective recovery of nutrients and pathogen removal. The anaerobic pre-treatment will remove high percent of the organic matter and will hydrolyze the nutrients present in the wastewater. In this form the design of the duckweed ponds can be based on nitrogen or pathogen removal and on the recovery of nitrogen in the form of biomass.

Nitrogen transformations and removal mechanisms in duckweed ponds

Molecular nitrogen (N_2) makes up almost 80% of the earth's atmosphere. Despite the abundance of nitrogen as a molecular gas, no eukaryote is able to make direct use of it. Instead nitrogen must be fixed combined with other elements, such as oxygen or hydrogen. The ability to fix atmospheric nitrogen is restricted to a limited number of bacteria and symbiotic associations between such bacteria and plants. The resulting compounds like the nitrate ion or ammonium ion are then used by autotrophic organism to produce more complex compounds like amino acids and proteins which are then used by plants and animals for incorporation into cellular biomass. The nitrogen cycle is one of the most complex biogeochemical cycles, because of the different oxidation-reduction states of nitrogen in nature from -3 in ammonia to +5 in nitrates.

Domestic wastewater contains organic nitrogen in the form of proteins, amino acids and other organic compounds, and inorganic nitrogen mainly as ammonium and small amounts of nitrogen oxides. In a wastewater treatment system these compounds are transformed and removed through different processes (Fig. 5). Also in duckweed ponds a set of microbial activities takes place and links these compounds to the complex nitrogen cycle (Fig. 6). Nitrogen concentrations are affected by a wide and diverse population of aerobic and anaerobic microorganisms, phytoplankton, zooplankton and duckweed. (Metcalf and Eddy, 1991). The oxygen balance in the system is very important for the nitrogen transformations and removal mechanisms. Below the most important transformations of nitrogen occurring in duckweed stabilisation ponds are briefly described.

Ammonification.
Organic compounds containing bound nitrogen present in the wastewater will be biodegraded by micro-organisms. In the degradation process, the proteins are split into peptides and amino acids. In the following step ammonium is set free (Fig. 5). Part of the organic matter will be degraded easily and will release ammonium very rapidly. Another part will be degraded slowly, in particular solids, which settle to the bottom of the system, and will be degraded through anaerobic mechanisms.

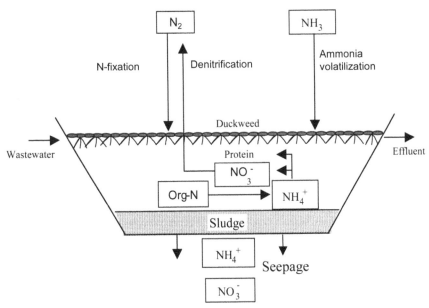

Fig. 5. Nitrogen transport and transformation mechanisms in a Duckweed Pond.

Ammonia volatilization.
Ammonium-nitrogen in water may be present as un-ionised gaseous form NH_3, or as NH_4^+ ion. The relative concentration of each species depends on the pH and temperature, where concentration of NH_3 (g) rises with increasing pH and temperature. This un-ionised form is rather volatile and will partly disappear into the atmosphere. Besides the total concentration of ammonium nitrogen, pH and temperature, the amount of volatized nitrogen depends also on the surface area in contact with the atmosphere and on the water depth.

Much of the nitrogen being removed in conventional stabilisation ponds is reported to be lost through the volatilization mechanism during the periods of high temperature and elevated pH (Pano and Middlebrooks, 1982; Azov and Tregubova, 1993; Silva *et al.*, 1995; Soares *et al.*, 1996). Elevated pH values are due to the photosynthetic activity of algae. In a duckweed pond, it is important to determine how important this nitrogen removal mechanism is because pH variation is expected to be less pronounced as algae populations in the water column are reduced considerably by shading of the duckweed cover. Zimmo *et al.* (2003b) found that ammonia volatilization was not an important nitrogen removal mechanism in duckweed ponds.

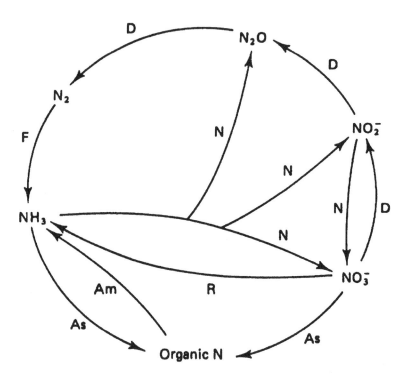

Fig 6. Different steps of nitrogen cycle as proposed by Robertson (1988).For simplicity, many intermediates have been omitted. As-assimilation; Am- ammonification; N-nitrification; F-nitrogen fixation; R-dissimilatory nitrate reduction.

Nitrification
The term nitrification is typically applied for the biological oxidation of ammonium nitrogen (NH_4^+-N) to nitrite (NO_2^--N) and further oxidation to nitrate (NO_3^--N):

$$NH_3 + 2 O_2 \xrightarrow{\text{Nitrifying bacteria}} NO_3^- + H_2O + H^+$$

It is generally assumed that autotrophic bacteria (*Nitrosomonas* and *Nitrobacter*), which are aerobic and chemolithoautotrophs, carry out this process. Based on stoichiometry, to oxidise 1 g of NH_4^+-N, 4.57 g of oxygen, and 7.14 g of alkalinity (as Calcium carbonate) are needed. Autotrophic nitrification is affected by factors like temperature, pH and oxygen concentrations. Optimum temperature range is 25-35 °C; optimal pH range is 7 - 8. Values of the half saturation coefficient for nitrification have been reported to range from 0.15-2 mg l^{-1} (Water Environment Federation, 1998)

The existence of heterotrophic nitrifiers has also been proved. Verstraete and Alexander (1973) showed evidence that heterotrophic nitrification may take place under certain conditions in nature. These findings raised the possibility that under appropriate conditions of pH, carbon and nitrogen supply, heterotrophic nitrification may take place and give rise to the formation of inorganic and organic nitrogenous products, though probably its contribution to nitrification is usually small. These results were supported later by Bergerova (1975). Heterotrophic nitrifying bacteria are better adapted to low oxygen concentrations. Therefore, the autotrophic nitrifying bacteria may loose the competition for oxygen at the sediment-water interface to bacteria that are better adapted. However, in spite of these findings on heterotrophic nitrification, Henriksen and Kemp (1988, as cited by Luijn, 1997) stated that autotrophic nitrification play a more important role in natural environments.

It has been frequently stated in literature that nitrification does not occur consistently in stabilization ponds due to the continuous environmental changes (Stone et al., 1975; Ferrara and Avci, 1982; Azov and Tergubova, 1993). Zimmo et al. (2003b), however, found that nitrification played an important role in nitrogen transformation of algae ponds. In duckweed ponds the environmental conditions are expected to be more stable than in algae ponds (Zimmo et al., 2002). The oxygen levels in the system will depend on factors like organic and nitrogen loads, retention time, pond depth, presence of pre-treatment and zones devoid of plants. Eighmy and Bishop (1989) studied the distribution and role of bacterial nitrifying populations in nitrogen removal in aquatic treatment systems and found that nitrifier populations were present and active year-round and equally distributed between the water column, macrophytes and sediments.

Denitrification.

$$NO_3^- \; + \; \text{Organic matter} \quad \xrightarrow{\text{Denitrifying bacteria}} \quad N_2$$

Denitrification is the dissimilatory reduction of oxidised nitrogen into gaseous oxides as intermediate products and gaseous nitrogen as final product. Most of the denitrifiers are facultative meaning they can use either oxygen or oxidized nitrogen as the terminal electron acceptor in respiration. In the absence of oxygen, oxidized nitrogen (nitrite-nitrate) is used as electro acceptor. Denitrification occurs mainly in anaerobic- anoxic conditions and with the availability of organic matter (Christensen and Tiedje, 1988). Dentrifying bacteria grow well in a pH range 7 – 8 and

15

temperature range between 5 and 25 °C, conditions typically found in wastewater. The presence of oxygen inhibits the activity of the denitrifying enzymes and suppresses their synthesis. From this, it could be concluded that if oxygen is available no NO_x^- is used, and that in the absence of oxygen nitrogen oxides act as terminal acceptors for the respiration (Knowles, 1982). This affirmation is supported by several observations. Erickson et al. (1996) who studied the denitrification capacity of epiphytic microbial communities in nutrient rich freshwater ecosystems never observed denitrification during illumination periods. They concluded that photosynthetic O_2 production probably exceeded heterotrophic O_2 respiration, inhibiting denitrification in the light. Christensen et al. (1990) studied the denitrification process in the sediments of a lowland stream and found that denitrification activity was reduced by 85% under light conditions when photosynthetic O_2 was produced. However, some researchers have demonstrated that some species of bacteria are available to denitrify under aerobic conditions (Robertson et al. (1988). Some anaerobic bacteria can perform a dissimilatory reaction to reduce nitrate to ammonium (Fig. 6). A competition between this process and denitrification for nitrate and organic matter may take place. The relative importance of the reduction of nitrate to ammonium is largely unknown (Robertson, 1988).

In stabilization ponds, nitrogen removal has often been attributed to ammonia volatilization (Pano and Middlebrooks, 1982; Azov and Tregubova, 1993; Silva et al., 1995; Soares et al., 1996) or to sedimentation (Ferrara and Avci, 1982). In recent studies with algae and duckweed pond, Zimmo et al. (2003b) found that denitrification was one of the most important removal mechanisms in both algae and duckweed ponds.

Anaerobic ammonium oxidation (ANAMOX).

$$NH_4^+ + NO_2^- \longrightarrow N_2 + 2H_2O$$

The ANAMOX process was recently discovered by Mulder et al. (1995). In this process nitrite or nitrate via nitrite is converted to dinitrogen gas under anaerobic conditions with ammonium as electron donor. The process is performed by catalyzing autotrophic bacteria that can convert nitrite to dinitrogen gas without the use of organic matter (Jetten et al., 1997).

Nitrogen fixation.
Although in some studies nitrogen-fixing activities in aquatic plant systems have been reported (Finke and Seeley (1978), Zuberer (1982)), nitrogen fixation is probably negligible in duckweed systems. The main reason for this observation is the repression of nitrogen-fixation by ammonium, a compound usually present in pond water (Brock et al., 1991).

Nitrogen compounds entering via deposition.
Some nitrogen compounds are present in low concentrations in the atmosphere, like nitrate, nitrite and ammonia. These may enter the duckweed pond via wet or dry deposition. It is assumed that this contribution will be minimal in the overall nitrogen balances of the pond.

Die off of biomass and settling to the bottom.
In duckweed ponds there will be high biomass production, mainly in the form of duckweed. Some of this biomass may die and settle to the bottom, where it will be degraded and nitrogen will be released into the water in the form of ammonium. This nitrogen flux is expected to be small if continuous harvesting is performed, since the percentage of decaying biomass is low at low or medium duckweed densities.

Percolation and seepage of water.
Percolation and seepage are important factors in the overall nitrogen balance of the system, if water loss through these mechanisms is considerable, because soluble nitrogen forms will be lost too. This will depend on the type of soil present in the area where the pond is placed, or the type of materials used in the pond construction.

Nitrogen up-take by duckweed.
Duckweed prefers ammonium nitrogen over other sources of nitrogen. This finding was reported by Ferguson and Bollard (1969) and Porath and Pollock (1982), who observed a rather efficient preferential ammonium uptake in the presence of nitrate in axenic (sterile) experiments with *Lemna gibba*. Although ammonium-nitrogen is the preferred form of nitrogen for duckweed growth, it may have inhibitory effects depending on the pH of the water. The rapid hydrolysis of organic nitrogen into the water may play an important role in nitrogen uptake by the plants and as a consequence on biomass production (Alaerts *et al.*, 1996). Values of daily nitrogen uptake by duckweed are presented in Table 3. The wide variation in the data presented in table 3 could be explained by the differences in the conditions present in each experience like duckweed species, water composition, nutrient concentration, temperature, harvesting regimes and biomass density.

Scope of this thesis
This thesis presents research on the effects of different operational variables on the performance of duckweed pond systems for domestic wastewater treatment. Operational variables included (1) the presence/absence of anaerobic pre-treatment, (2) the insertion of regular algae ponds into a series of duckweed ponds to create zones with higher oxygen concentrations and (3) pond depth. Special attention was paid to the fate of nitrogen in the system, since nitrogen is a 'pollutant' that is removed by various processes, but also recovered via duckweed growth. Nitrogen in wastewater can be transformed into a new resource. So far, only few studies have been undertaken to evaluate operational duckweed pond schemes. Therefore design criteria for this technology-under-development are still lacking.

Table 3 Nitrogen and phosphorus uptake (g m^{-2} d^{-1}) adapted from Gijzen and Kondker (1997).

Region	Species	Nitrogen	Phosphorus	Reference
Florida	S. polyrrhiza	-	0.015	Sutton and Ornes(1977)
Louisiana	Duckweed	0.47	0.16	Culley et al. (1978)
CSSR	Duckweed	0.2	-	Kvet et al. (1979)
Italy	L.gibba.L.minor	0.42	0.01	Corradi et al.(1981)
USA	Lemna	1.67	0.22	Zirschky and Reed (1988)
India	Lemna	0.50-0.59	0.14-0.30	Tripathi et al. (1991)
Minnesota	Lemna	0.27	0.04	Lemna Corporation
Bangladesh	S.polyrrhiza	0.26	0.05	Alaerts et al. (1996)
Yemen	Lemna gibba	0.05-0.2	0.01-0.05	Al-Nozaily (2001)
Palestine	Lemna gibba	0.2-0.55	-	Zimmo et al. (2003b)
Egypt	Lemna gibba	0.44	0.092	El-Shafai et al.(2004)

Chapter 2 reports on a study on the effect of anaerobic pre-treatment on the environmental and physico-chemical conditions of duckweed based stabilization ponds. These environmental and physico-chemical conditions affect both plant growth and biological processes in the system and as a consequence the performance in terms of removal efficiencies and biomass production. In the anaerobic pre-treatment, organic nitrogen is hydrolysed and converted into ammonium nitrogen. This is the preferred source of nitrogen for duckweed, but high concentrations may become inhibitory to the plant. The effect of total ammonium nitrogen concentration and pH on growth rates of duckweed are described in **Chapter 3**.

The effect of anaerobic pre-treatment on the overall performance of duckweed stabilisation ponds, in terms of organic matter, suspended solids, nitrogen, phosphorus, pathogen removal and biomass production, is presented in **Chapter 4**. **Chapter 5** reports on investigations into nitrogen transformation mechanisms and mass balances in duckweed stabilization ponds. Anaerobic pre-treatment significantly changes the environmental conditions that affect nitrogen transformation mechanisms. Therefore comparisons were performed between systems with and without anaerobic pre-treatment.

Oxygen concentration affects nitrogen transformations and removal via nitrification and denitrification in the duckweed system. Therefore the effect of the introduction of aerobic zones by removing the duckweed cover from some of the ponds at an early stage was studied and the results are presented in **Chapter 6**. Pond depth affects not only oxygen concentrations, but also the contact between plants and water column, hydraulic retention times and loading rates. Therefore the effect of this parameter on the nitrogen balance and removal mechanisms was studied and described in **Chapter 7**.

Duckweed ponds are modified conventional stabilisation ponds with the water surface covered with a duckweed mat. The mat is expected to trigger considerable changes in the pond environment and therefore influences the different transformations and processes. A comparison of the performance between a duckweed stabilisation pond and a conventional stabilisation pond was performed at full-scale and the results are presented in **Chapter 8**. Finally a summary of the main discussions, recommendations and conclusions is presented. .

References
Alaerts G., Mahbubar Rahman, Kelderman P. (1996). Performance Analysis of a full-scale duckweed-covered sewage lagoon. *Wat. Res.* 30 (4), 843-852.

Al-Nozaily F. A. (2001). Waste stabilization pond performance in Sana'a, Yemen and the influence of purple non-sulphur bacteria. In: Performance and Process Analysis of Duckweed-Covered Sewage Lagoons for high Strength Sewage. Doctoral Dissertation. Delft University of Technology- International Institute of Hydraulic and Environmental Engineering. Delft-The Netherlands.

Araujo M. C. (1987). Use of water hyacinth in tertiary treatment of domestic wastewater. *Wat. Sci. Tech.* 19, 11-17.

Azov Y. and Tregubova T. (1993). Nitrification processes in stabilisation reservoirs. *Proceedings of the 2nd IAWQ International Specialist Conference on waste stabilisation ponds and the reuse of pond effluents*. California, Oakland.

Bergerova E. (1975). Nitrification by heterotrophs. *Folia Microbiologica.* 20 (1), 77.

Brix H. and Schierup H. H. (1989). The use of aquatic macrophytes in water pollution control. *Ambio.* 18, 100-107.

Brock T. D., Madigan M., Martinko J., Parker J. (1991). Biology of Microorganism. 17th Edition. Prentice Hall Inc., London.

Caicedo J.R., Espinosa C., Gijzen H., Andrade M. (2002). Effect of anaerobic pre-treatment on physicochemical and environmental characteristics of Duckweed based ponds. *Wat. Sci.Tech.* 45(1), 83-89.

CEPIS-0PS-0MS. (2004). Proyecto regional sistemas integrados de tratamiento y uso de agues residuals en America Latina. Realidad y potencial. Convenio IDRC-

OPS/HEP/CEPIS 2000-2002. http://www.cepis.ops-oms.org. Consulted in April 2004.

Christensen S. and Tiedje J. M. (1988). Oxygen control prevents denitrifiers and barley roots from directly competing for nitrate. *FEMS Microbiol. Ecol.* 53, 217-221.

Christensen P. B., Nielsen L. P., Sorensen J., Revbech N. P. (1990). Denitrification in nitrate-rich streams: Diurnal and seasonal variation related to benthic oxygen metabolism. *Limnol. Oceanogr.* 35 (3), 640-651.

Clark J. R., Hassel J. H. van, Nicholson R. B., Cherry D. S., Cairns J. (1981). Accumulation and depuration of metal by Duckweed (*Lemna perpusilla*). *Ecotoxicology and Environmental safety.* 5, 87-96.

Corradi M, Copelli M., Ghetti P. (1981). Olture di Lemna su scarichi zootecnici. Inquinamento, 23, 45-49.

Cosgrove W.J. and Rijsberman F. R. (2000). World water vision, making water every's body's business. World Water Council. Earthsean Publications Ltd. London.

Culley D. D. and Epps E. A. (1973) Use of duckweed for waste treatment and animal feed. *Jour.Wat.Poll. Cont. Fed.* 45(2), 337-347.

Culley D. D., Gholson J. H., Chisholm T. S., Standifer L. C., Epps E. A. (1978). Water quality renovation of animal waste lagoons utilizing aquatic plants. U.S. Environmental Protection Agency. Ada. Oklahoma. pp 166.

Edwards P., Pacharaprakiti C., Yomjinda M. (1990). Direct and indirect reuse of septage for culture of Nile Tilapia *Oreochromis niloticus*. In Proceedings of The second Asian Fisheries Forum. Asian Fisheries Society. Hirano R. and Hanyu I., (eds.), Manila, Philippines.

El-Shafai S. A., El-Gohary F., Nasr F.A., Van der Steen N. P., Gijzen H. J. (2004). Nutrient recovery from domestic wastewater using a UASB-duckweed pond system. Submitted to *Bioresource Technology*

Eriksson P. G. and Weisner S. E. B. (1996). Functional differences in epiphytic microbial communities in nutrient-rich freshwater ecosystems: an assay of denitrifying capacity. *Freshwater Biology.* 36, 555-562.

Ferguson A. R. and Bollard E. G. (1969). Nitrogen Metabolism of *Spirodela oligorrhiza*. I. Utilisation of ammonium, nitrate and nitrite. *Planta* (Berl.). 88, 344-352.

Ferrara R. A. and Avci C. B. (1982). Nitrogen dynamics in waste stabilisation ponds. *Jour. Wat. Poll. Cont. Fed.* 54(4), 361-369.

Finke L. R. and Seeley H. W. (1978). Nitrogen fixation (acetylene reduction) by epiphytes of freshwater macrophytes. *Applied and Environmental Microbiology*, 36 (1), 1129-138.

Gijzen H. J. and Khondker M. (1997). An overview of the ecology, physiology, cultivation, and application of Duckweed, Literature Review. Report of Duckweed Research Project. Dhaka, Bangladesh.

Gijzen H.J. and Ikramullah M. (1999). Pre-feasibility of duckweed-based wastewater treatment and resource recovery in Bangladesh. World Bank Report, Washington D. C.

Gijzen H. J. (2001). Anaerobes, aerobes, and phototrophs: a wining team for wastewater management. *Wat. Sci Tech*. 44(8), 123-132.

Gijzen H. J. (2002). Anaerobic digestion for sustainable development: a natural approach. *Wat. Sci. Tech.*, 45(10), 321-328.

Gijzen H. J. (2004). A 3-step strategic approach to sewage management for sustainaible water resources proteccion. *Wat, Sci. Tech*. In press.

Graaf J. H. J. M van der, Meester-Broertjes H. A., Bruggeman W. A., Vles E. J. (1997). Wat. Sci. Tech. 35 (10), 213-220.

Green F. B., Bernstone L. S., Lundquist T. J., Oswald W. J. (1996). Advanced integrated wastewater pond systems for nitrogen removal. *Wat. Sci. Tech*. 33 (7), 207-217.

Haandel A. van and Lettinga G. (1994). Anaerobic Sewage Treatment. A practical guide for regions with a hot climate. John Wiley & Sons. New York.

Haandel A. van and Catunda P. F. C. (1997). Anaerobic digestion of municipal sewage and post-treatment in stabilisation ponds. Memories of International Conference. Mazatlan, Mexico

Harvey R. M. and Fox J. L. (1973). Nutrient removal using *Lemna minor*. *Jour. Wat. Poll. Cont. Fed.* 45(9), 1928-1938.

Ice J. and Couch R. (1987) Nutrient absorption by Duckweed. *J. Aquat. Plant Manage.* 25, 30-31.

Jetten M.S. M., Horn S.J., Van Loosdrecht M. C. M. (1997). Towards a more sustainable municipal wastewater treatment system. Wat. Sci. Tech. 35 (9), 171-180.

Kawabata K. and Tatsukawa R. (1986) Growth of duckweed and nutrient removal in a paddy field irrigated with sewage effluent. *J. Environmental Studies.* 27, 277-285.

Knowles R. (1982). Denitrification. *Microbiol. Reviews.* 46, 43-70.

Kvet J., Rejmankova E., Rejmanek M. (1979). Higher aquatic plants and biological wastewater treatment. The outline of possibilities. Aktiv Jihoceskych Vodoh Conf. pp 9.

Landolt E. (1986). The family of *Lemnaceae*-a monographic study, Vol.1. *Veroeffentlichungen des geobotanisches Institutes der ETH*, Stiftung Rubel, 71, Zurich, 1-566.

Landolt E and Kandeler R. (1987). The family of *Lemnaceae* monographic study, Vol.2. *Veroeffentlichungen des geobotanisches Institutes der ETH*, Stiftung Rubel, 95, Zurich, 1-638.

Larsen T. A. and Gujer W. (1997). The concept of sustainable urban water management. *Wat. Sci. Tech.* 38 (9), 1-10.

Luijn F. van (1997). Nitrogen removal by denitrification in the sediments of a shallow lake. Ph. D. Thesis. University of Wageningen. Wageningen. The Netherlands

Mbagwu I. G. and Adeniji H. A. (1988). The nutritional content of duckweed (*Lemna puntata*) in the Kainji Lake area, Nigeria. *Aquat. Bot.*, 29, 357-366.

McLay C. L. (1976). The effect of pH on the population growth of three species of duckweed: *Spirodela oligorrhiza, Lemna minor* and *Wolffia arrhiza. Freshwater Biology.* 6, 125-136.

Metcalf and Eddy. (1991). Waste Engineering. Treatment, disposal and reuse. Tchobanoglous G. and Burton F. L. [eds.]. 2nd Ed. McGraw Hill, Inc., USA.

Mulder A., Graaf A. A., Robertson L. A., Kuenen J. G. (1995). Anaerobic ammonium oxidation discovered in a denitrifying fluidized bed reactor. FEMS Microbiol. Lett. 16, 177-184.

Oron G., Wildschut L. R., Porath D. (1985). Waste water recycling by duckweed for protein production and effluent renovation. *Wat. Sci. Tech.* 17, 803-817.

Oron G. (1994). Duckweed culture for wastewater renovation and biomass production. *Agricultural Water Management.* 26, 27-40.

Pano A. and Middlebrooks E. J. (1982). Ammonia nitrogen removal in facultative wastewater stabilisation ponds. *Jour. Wat. Poll. Cont. Fed.* 54(4), 344-351.

Porath D. and Pollock J. (1982). Ammonia stripping by duckweed and its feasibility in circulating aquaculture. *Aquat. Bot.* 13, 125-131.

Rao S. V. R. (1986). A review of the technological feasibility of aquaculture for municipal wastewater treatment. *Intern. J. Environmental Studies.* 219-223.

Reddy K. R. and Smith W. H. (Eds.). (1987). Aquatic plants for water treatment and resource recovery. Magnolia Publishing Inc. Orlando-Florida.

Reed S. C., Middlebrooks E. J., Crites R. W. (1995). Natural systems for waste management and treatment. 2nd Ed. Mc Graw Hill. New York.

Robertson L. A. (1988). Aerobic denitrification and heterotrophic nitrification in *Thiosphaera Pantotropia* and other bacteria. Ph. D. Dissertation. Delft Technical University. The Netherlands.

Selvam L. P., Shamsuddin A. J., Ikramullah M., Mudgal A. K. (1992). *Lemnaceae* based wastewater treatment and resource recovery. Proceedings 18th Conference Water, Environment and Management. Kathmandu, Nepal.

Skillicorn P., Spira W., Journey W. (1993). Duckweed aquaculture, a new aquatic farming system for developing countries. The World Bank. 76 p. Washington.

Silva S. A., de Olivieira R., Soares J., Mara D. D., Pearson H. W. (1995). Nitrogen removal in pond systems with different configurations and geometries. *Wat. Sci. Tech.* 31 (120), 321-330.

Smith S. and Kwan M. K. H. (1989). Use of aquatic macrophytes as a bioassay method to assess relative toxicity, uptake kinetics and accumulated forms of trace metals. *Hydrobiologia.* 188/189, 345-351.

Soares J., Silva S. A., de Oliveira R., Araujo A. L. C., Mara D. D., Pearson H. W. (1996). Ammonia removal in pilot scale WSP complex in Northeast Brazil. *Wat. Sci. Tech.*, 33 (7), 165-171.

Steen N. P. van der, Brenner A., Oron G. (1998). An integrated duckweed algae pond system for nitrogen removal and renovation. *Wat. Sci. Tech.* 38(1), 335-343.

Steen N. P. van der, Brenner A., Buuren J. van, Oron G. (1999). Post-treatment of UASB reactor effluent in an integrated duckweed and stabilization pond system. *Wat. Res.* 33(3), 615-620

Steen N. P. van der, Nakiboneka P., Mangalika L., Ferrer A. V. M., Gijzen H. J. (2003). Effect of dukweed cover on greenhouse gas emissions and odour release from waste stabilization ponds. Wat. Sci. Tech. 48 (2), 341-348.

Stone R. W., Parker D. S., Cotteral J. A. (1975). Upgrading lagoon effluent for best practicable treatment. Jour. Wat. Poll. Cont. Fed. 47 (8), 2019-2042

Sutton D. L., Ornes W. H. (1977). Growth of Spirodela polyrhiza in static sewage effluent. *Aquatic Botanic*, 3, 231-237.

Tripathi B. D., Srivastava J., Misra K. (1991). Nitrogen and phosphorus renoval capacity of four choosen aquatic macrophytes in tropical freshwater ponds. J. *Env. Qual.*, 18, 143-147.

Ullrich W. R., Larsson M., Larsson C. M., Lesch M., Novacky A. (1984). Ammonium uptake in *Lemma gibba* G 1, related membrane potential charges, and inhibition of anion uptake. *Physiol. Plant, Copenhagen,* 61, 369-376.

Verstraete W. and Alexander M. (1973). Heterotrophic nitrification in samples of natural ecosystems. *Env. Sci. Tech.*, 7 (1), 39-42.

Wang W. (1986). Toxicity test of aquatic pollutants by using common duckweed. *Environmental Pollution* (Series B), 11, 1-14.

Wang W. (1990). Literature review on duckweed toxicity testing. *Env.Res.* 52, 7-22.

Water Environment Federation (1998). Biological and chemical systems for nutrient removal. Alexandria, USA.

Zimmo O., Al-Sa'ed R. M., Van der Steen N. P, Gijzen H. J. (2002). Process Performance Assessment of algae-based and duckweed-based wastewater treatment systems. *Wat. Sci. Tech.*, 45(1), 9i-101.

Zimmo O., Van der Steen N. P, Gijzen H. J. (2003a). Effect of organic surface load on process performance of pilot scale algae and duckweed-based waste stabilization ponds. In: Nitrogen transformation and removal mechanism in algal and duckweed waste stabilization ponds. Ph. D. Dissertation. Wageningen University and International Institute of Hydraulic and Environmental Engineering. The Netherlands.

Zimmo O., Van der Steen N. P, Gijzen H. J. (2003b). Nitrogen mass balance over pilot scale algae and duckweed-based wastewater stabilization ponds. In: Nitrogen transformation and removal mechanism in algal and duckweed waste stabilization ponds. Ph. D. Dissertation. Wageningen University and International Institute of Hydraulic and Environmental Engineering. Holland

Zirschky J. and Reed S. (1988). The use of duckweed for wastewater treatment. *Jour. of Wat. Poll. Cont. Fed.* 60 (7), 1253-1258.

Zuberer D. A. (1982). Nitrogen fixation (acetylene reduction) associated with duckweed (*Lemnaceae*) mats. *Applied and Environmental Microbiology*, 43 (4), 823-828.

Chapter 2

Effect of anaerobic pre-treatment on environmental and physicochemical characteristics of duckweed stabilization ponds.

Adapted from:

Caicedo J.R., Espinosa C., Gijzen H., Andrade M. (2002). Effect of anaerobic pre-treatment on physicochemical and environmental characteristics of Duckweed based ponds. *Wat. Sci.Tech.* 45(1), 83-89.

Chapter 2

Effect of anaerobic pre-treatment on environmental and physicochemical characteristics of duckweed stabilization ponds.

Abstract.

Duckweed stabilization ponds, as an alternative for wastewater treatment, are attracting a growing interest from researchers because they are low cost, easy to built and operate and produce tertiary quality effluents. Besides, this technology offers the possibility of resource recovery by producing high quality duckweed protein, which can be of further use. The presence of a layer of duckweed on the surface of the ponds is expected to produce different environmental and physicochemical conditions in the water compared to those in conventional stabilization ponds. The environmental and physicochemical conditions affect both plant growth and microbiological treatment processes in the system. Two series of continuous flow pilot plants, composed of seven ponds in series each, were operated side by side. One system received artificial sewage with anaerobic pretreatment, while the other system received the same wastewater without anaerobic pretreatment. The flow was kept constant during the operation. pH, temperature, dissolved oxygen, alkalinity, conductivity, biochemical oxygen demand, total and ammonium nitrogen, nitrites and nitrates, phosphorus were monitored under steady state conditions. The main conclusions from this study include the following: pH levels are very stable in both systems with and without anaerobic pretreatment. Temperature gradients are present during daytime but not as high as they may be in conventional stabilization ponds. Oxygen levels are significantly higher in the duckweed system with anaerobic pretreatment, especially in the top layer. Nevertheless, re-aeration rates are low in both systems. Both systems were efficient in removing organic matter. The system without pretreatment obtained 98% of BOD_5 removal in pond 4, so 12 days of retention time will be enough to reach high organic matter removal. The system with pretreatment obtained also 98% BOD_5 removal (92% of BOD_5 in UASB reactor). In this case with a very efficient UASB reactor, the duckweed ponds will serve as a polishing step for remaining organic matter. Nutrient removals were 37-48% for nitrogen and 45-50 % for phosphorus in the lines with and without pretreatment respectively. It is important to establish the nitrogen balance for the systems in order to generate a better understanding of the nitrogen transformations in the duckweed system.

Key words

Anaerobic effluent post treatment, duckweed, duckweed pond pretreatment, *Lemnaceae,* stabilization ponds, *Spirodela polyrrhiza*, wastewater treatment.

Introduction

Urbanization as a result of population growth, industrialization and agricultural activities is causing increasing environmental pollution problems. The conventional processes for wastewater treatment including biological and chemical processes to remove unwanted substances and organisms from wastewater are costly in terms of investment and operation and many countries can not afford to build and operate such systems.

The development of new effective technologies with high treatment efficiencies and low construction and operation costs is required to address the increasing wastewater problems in developing regions (Zirscky and Reed, 1988; Brix and Schierup, 1989). During recent years much research has gone into land applications systems, constructed wetlands, algae ponds, and aquatic plants systems (Oron et al., 1986; Skillicorn et al., 1993; Reed et al., 1995; Alaerts et al., 1996)). The use of floating aquatic plants like duckweed in stabilization ponds seems to be attractive if combined with re-use schemes for the produced plant biomass (Gijzen and Kondker, 1997; Gijzen and Ikramullah, 1999; Gijzen, 2001).

The biological and physicochemical processes occurring in conventional stabilization ponds are complex. Depending on the organic loading, aerobic, anaerobic or facultative zones can be present (Metcalf and Eddy, 1991; IMTA, 1992). The presence of a duckweed cover on a stabilization pond introduces even more complexity to the system because of interactions between plants and water and reduction of light penetration. This is expected to cause major changes on the environmental and physicochemical characteristics in the water column. The environmental conditions prevailing in conventional stabilization ponds have been described, but for macrophyte-based ponds this has not been studied in detail. A good understanding of the environmental conditions is important since these will affect the microbiological reactions involved in the treatment processes in stabilization ponds.

One limitation which has been mentioned in literature for duckweed ponds is the reduced oxygen levels in the water column, and consequently lower organic matter removal capacity, compared to conventional ponds (Reed et al., 1995). As a consequence lower organic load can be applied, while larger surface area would be required. In an attempt to reduce the area requirement we have studied the effect of anaerobic pre-treatment, prior to duckweed stabilization ponds. Anaerobic pretreatment may also change environmental and physicochemical characteristics in the ponds, as the organic matter will be greatly reduced in the anaerobic reactor. Besides, due to hydrolysis of organic matter, nutrients may be present in a soluble form ready to be used by the plants.

As the pathways of contaminants degradation depend very much on environmental conditions and on the physicochemical equilibrium in the water, it is important to study the conditions that are established in duckweed ponds in order to understand the different processes involved in the removal of specific contaminants.

Methodology
Experimental set up

The pilot plant consists of two series of ponds operated in parallel. Both series consist of seven ponds. One treatment line is preceded by a UASB reactor (Fig. 1). The system is located outdoors at the university campus, exposed to natural conditions but protected from rainwater when necessary (Fig. 2). The minimum, average and maximum air temperatures during the experimental period were 19.2 °C, 23.2 °C and 29.3 °C respectively. The minimum, average and maximum solar radiation was 378, 413 and 446 cal m^{-2} d^{-1} respectively.

Each line was operated in parallel under continuous flow and similar conditions. The UASB reactor has a volume of 23 l, a diameter of 0.15 m and sludge depth 0.40 m. The duckweed ponds have a volume of 170 l each, with a depth of 0.70 m and a diameter of 0.56 m. The duckweed species used was *Spirodela polyrrhiza*

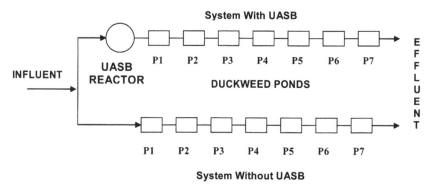

Fig. 1. Schematic diagram of pilot scale continuous flow plant

Fig 2. Pilot scale continuous flow plant

Artificial Wastewater Composition
Synthetic wastewater was prepared to simulate the characteristics of domestic wastewater. The composition (on COD basis) was as follows: protein 50%, sugar 8%, cellulose 8%, oil and detergent 10%, starch 24%. The total chemical oxygen demand was around 400 mg l^{-1}, which corresponds to medium strength wastewater (Metcalf and Eddy, 1991). Total nitrogen concentration (TKN) was in the range of 30-40 mg l^{-1} while phosphorus was adjusted to 3.6-3.8 mg l^{-1}. The artificial wastewater was prepared in tap water. Macro and micronutrients were added as indicated in Table 1.

Operational conditions
The UASB reactor was operated at a hydraulic retention time (HRT) of 10 hours. Each line of the pilot plant was fed with the synthetic waste with a flow of 2.3 l h^{-1}, resulting in a HRT of 3 days per pond. The duckweed cover was harvested every four days from each pond to leave a biomass density of 700 g m^{-2} (fresh weight) after each harvesting. This density produced a full cover to prevent light penetration and growth of algae.

Every harvesting period, temperature, pH and oxygen concentrations were determined for each pond at three different depths (7, 35 and 63 cm from the surface) and always between 11 a. m. and 3 p.m. Grab samples were collected from the influent wastewater, from the effluent of the UASB reactor and from the effluent of each pond. Samples were analyzed for alkalinity, conductivity, chemical oxygen demand, and biochemical oxygen demand. Nutrient levels (nitrogen and phosphorus) were also analyzed.

Table 1. Macro and micronutrients.

Macro nutrients	mg l^{-1}
Urea	42.9
K_2HPO_4	11.9
KH_2PO_4	8.8
$MgCl_2.6H_2O$	7.0
NaCl	40.0
Micro nutrients	**mg l^{-1}**
EDTA	13.3
$FeCl_3.6H_2O$	4.4
$MnSO_4.H_2O$	0.09
$CoSO_4.7H_2O$	0.03
$ZnSO_4.7H_2O$	0.03
H_3BO_3	0.01
$(NH_4)8Mo_7.O_{24}.4H_2O$	0.02
$Na_2SeO_3.5H_2O$	0.03
$NiCl_2.6H_2O$	0.12
$CuSO_4$	0.04

Analytical methods
Biochemical oxygen demand (BOD_5), chemical oxygen demand (COD), total suspended solids (TSS), alkalinity, total Kjeldahl nitrogen (TKN), ammonium nitrogen (NH_4^+-N), nitrites (NO_2-N) and nitrates (NO_3-N) were measured according to Standard Methods (APHA, 1995). Dissolved oxygen, pH, temperature and conductivity were measured with electrodes.

Just before duckweed harvesting, the biomass density was determined to calculate the amount to be harvest to leave a density of 700 g m^{-2} (fresh weight). Dry matter was determined by drying the plants in the oven at 70°C during 24 hours.

Results for different treatment lines were compared using ANOVA test and results for ponds within each treatment line were compared using the paired t-test. Confidence limit was 95%.

Results and discussion
The wastewater and UASB effluent characteristics are presented in Table 2. The results obtained in the duckweed pond systems are presented in Tables 3 and 4.

Table 2. Wastewater and UASB effluent characteristics

Parameters	Wastewater		UASB Effluent.	
	av.	s.e	av.	s.e
pH	7. 2	0.1	7.1	0.1
Temperature	25	0.9	25.1	1.0
Alkalinity (mg CaCO$_3$ l^{-1})	26.9	2.1	154.3	5.8
Conductivity (μ S cm^{-1})	293	1.7	559	5.4
Total nitrogen (mg N l^{-1})	36.2	1	36.2	1.1
Ammonia nitrogen (mg N l^{-1})	0.97	0.4	34.74	2
Total phosphorus (mg P l^{-1})	3.7	0.1	3.6	0.1
COD (mg l^{-1})	392	11.3	71	2.6
BOD$_5$ (mg l^{-1})	163	21.1	12.5	0.5

av. average;
se standard error

pH and alkalinity
The pH levels found in both systems were in the optimal range reported in literature for the growth of *Spirodela polyrrhiza* (Bitcover & Sieling, 1951; Landolt and Kandeler, 1987). The pH levels between the two systems were very similar with slightly lower values in the first two ponds of the system without pretreatment. In each system, pH was very stable through the different ponds. Low pH gradients were obtained in the vertical profiles (maximum average variation of 0.1 units). No significant differences were found with respect to depth. Some measurements taking during the night did no showed variation between day and night.

Table 3. Results of the different parameters measured in the system without UASB reactor

Parameters		PONDS												
		1		2		3		4		5		6		7	
		av.	s.e	av.	s.e	av.	s.e	av.	s.e	av.	s.e	av.	s.e	av.	s.e
*pH	7 cm	6.8	0	6.8	0	7	0	7	0	7	0	7	0	7	0
	35 cm	6.7	0	6.8	0	7	0	7	0	7	0	7	0	7	0
	63 cm	6.7	0	6.8	0	7	0	7	0	7	0	7	0	6.9	0
*Temperature	7 cm	27.4	1.1.	27.1	1	27	1	26.5	1	26.1	0.9	25.9	1.1	26.1	1.2
	35 cm	26.1	1.1	26	1	25.9	1	25.7	1	25.4	1	25.5	1	25.7	1.1
	63 cm	25.5	1	25.2	0.9	25.2	0.9	25	0.9	24.8	0.8	24.9	0.9	25.1	0.9
*Oxygen	7 cm	0.7	0.2	0.7	0.2	0.7	0.1	0.7	0.1	0.9	0.1	1.1	0.2	1.1	0.2
	35 cm	0.2	0	0.2	0.1	0.2	0	0.2	0	0.3	0.1	0.7	0.2	1.4	0.3
	63 cm	0.1	0	0.1	0	0.1	0	0.1	0	0.2	0	0.6	0.2	1.1	0.3
Alkalinity(mg l⁻¹)		141.5	1.1	145.4	1.5	142.7	0.9	130.9	3.6	122.4	1.5	110.6	2.1	99.4	1
Conductivity (µ s cm⁻¹)		531	2.8	529	3.4	515	3.2	499	9.2	482	6.7	456	8	432	6.9
Total nitrogen. (mg N l⁻¹)		35	1.1	31.7	1.2	28.7	1	28.7	0.8	23.1	1.4	19.8	1.1	18.7	1.1
Ammonia nitrogen (mg N l⁻¹)		31.9	0.9	30.8	0.8	28.2	0.8	24.8	0.8	23.6	0.5	21.5	0.7	18.7	0.7
Total p. (mg P l⁻¹)		3.4	0.1	3.3	0.1	3.1	0.1	2.9	0.1	2.4	0.1	2.1	0.1	1.9	0.1
COD (mg l⁻¹)		126	14	108	3.7	92	2.7	85	3.1	74	3.9	65	2.5	63	3.8
BOD (mg l⁻¹)		61.8	6.8	28.7	0.2	13.9	0.3	4.2	1.6	4.4	1.8	3.8	1.0	2.9	0.7

av =average
s.e = standard error
* measurements at different depths

Effect of Operational Variables on Nitrogen Transformations in Duckweed Stabilization Ponds

Table 4. Results of the different parameters measured in the system with UASB reactor

parameters		PONDS												
		1'		2'		3'		4'		5'		6'		7'
		av.	s.e	av.	s.e	av.	s.e	av.	s.e	av.	s.e	av.	s.e	av.	s.e
*pH	7 cm	6.9	0	7	0	7	0	7	0.1	7	0.1	7	0.1	6.9	0.1
	35 cm	6.9	0	7	0	7.1	0	7	0.1	7	0.1	7	0.1	6.9	0.1
	63 cm	6.9	0	7	0	7.1	0	7.1	0.1	7	0.1	7	0.1	6.9	0.1
*Temperature	7 cm	27.3	1.3	27.3	1.2	27.2	1.1	26.9	1	26.6	1.2	26.3	1.2	26.1	1.2
	35 cm	26.3	1.1	26.1	1.1	26.4	1.1	26.2	1	26	1.1	25.7	1.1	25.6	1.1
	63 cm	25.4	1	25.6	1	25.5	0.9	25.6	0.9	25.3	0.9	25.2	0.9	25	0.9
*Oxygen	7 cm	1.1	0.2	0.9	0.2	1.3	0.3	1.3	0.2	1.3	0.3	2	0.3	2	0.3
	35 cm	0.3	0.1	0.4	0.1	0.4	0.1	0.7	0.2	0.5	0.1	1.4	0.2	1.6	0.2
	63 cm	0.2	0	0.2	0	0.2	0.1	0.4	0.1	0.5	0.2	1.2	0.2	1.2	0.1
Alkalinity(mg l⁻¹)		169.5	2.5	164.4	2.4	153.9	2	142.2	4.3	127.2	4.9	108.7	5.7	88.5	6.7
Conductivity (µ s cm⁻¹)		578	5	571	4.6	557	7.7	527	6.6	501	8.5	470	10.4	434	11.4
Total nitrogen. (mg N l⁻¹)		35.3	1.1	32.6	0.8	30.9	0.7	29.8	0.7	26.8	1.3	21.8	1.2	22.9	0.5
Ammonia nitrogen (mg N l⁻¹)		34.9	1.3	33.5	0.6	31.3	0.8	28.8	0.7	25.4	0.9	22.4	1.2	18.7	1.2
Total p. (mg P l⁻¹)		3.3	0.2	3.3	0.1	3.2	0.1	2.9	0.1	2.7	0.1	2.3	0.1	2.0	0.1
COD (mg l⁻¹)		69	2.2	63	3.6	56	2.5	58	3.7	55	2.9	51	1.7	50	3
BOD (mg l⁻¹)		7.2	3.5	5.3	3.0	2.3	1.2	2.7	1.2	4	0.3	1.8	0.3	2.8	1.1

av =average

s.e = standard error

* measurements at different depths

The stability in pH may be partially explained by the levels of alkalinity. Although it decreased gradually in both systems, the remaining level was enough to keep buffer capacity intact up to the last pond. The low variation of pH may also suggest low algae activity in the ponds, which can be confirmed with the low oxygen levels and gradients that were present in the ponds during the period of the day when higher photosynthetic activity was expected.

Low variation of pH levels in duckweed ponds represent an important difference when compared with conventional stabilization ponds where pH fluctuations of several pH units may be observed during the day. Even though duckweed can grow in a wide range of pH (Landolt and Kandeler, 1987), a stable level may enhance biomass production in the system. Besides, many microbial populations would be favored with the steady pH conditions

Temperature
No significant differences were found between the treatments at the same depth, while significant differences were found between depths. A temperature gradient between top and bottom is formed during the day with a maximum average difference of $2^{\circ}C$.

It can be pointed out that even though temperature gradients were present in the duckweed ponds during day time, these are not as high as the ones reported for conventional stabilization ponds (Cubillos, 1994; Mendonca, 1999). This can be due to the presence of the plant cover, which absorbs and reflects the solar radiation reducing the warming up of the top water layers. During the night temperature differences in the water column are decreasing and by the morning, just before sunrise, no temperature gradient was observed anymore

The presence of low temperature gradients in duckweed pond reduces the possibility of water mixing by convection with the benefic for the sedimentation process occurring on the system. This may be one of the reasons for the high quality of duckweed effluents in terms of suspended solids.

Dissolved oxygen
Oxygen concentrations increase gradually with the retention time in each system. Higher levels of oxygen were established in the system with pretreatment where concentrations above 1 mg l^{-1} were observed in the top layer from pond 1 to 7, while in the system without pretreatment only ponds 6 and 7 reached more than 1 mg l^{-1} in the top layer. Except for pond 7 of the system without pretreatment, significant differences were found between the top layers and middle layer for all ponds of both systems. Middle and bottom layers were not significantly different for most of the ponds. Some measurements taken during the night showed a slight reduction of the oxygen concentrations in the ponds.

33

In spite of the significant differences, it was confirmed that oxygen concentrations in duckweed ponds are low compared to conventional ponds at day time (Reed *et al.*, 1995). This low oxygen levels reached in the systems denoted a low algae activity. The dense cover of duckweed plants may have reduced the direct oxygen transference from the atmosphere but it can transport it indirectly through the root zone (Moorhead and Reddy, 1988). Then the gradual oxygen increase along the systems was most probably due to duckweed matt oxygen transference.

Organic matter removal
In the system without pretreatment the surface organic loading rate was in the range of   364 – 8.5 Kg ha^{-1} d^{-1} between the first and last pond. The load to the first pond was slightly higher than the load recommended by Mara *et al.* (1992) for facultative ponds.   Most of this removal occurred between pond 1 and 4. Organic matter removal was 98 % in terms of BOD$_5$. Low oxygen levels were observed up to pond 4 and started to increase gradually up to pond 7. In spite of the low oxygen levels it was possible to obtain high level of organic matter removal at a hydraulic retention time of 12 days. This is an important observation as low organic matter removal has been reported to be one of the main limitations of duckweed pond technology due to low oxygen concentrations in the water column.

In the system with pretreatment, the organic loading rate was very low (28 - 4 Kg ha^{-1} d^{-1}) due to the high removal efficiency of the UASB reactor (92 % of BOD$_5$). Total BOD$_5$ removal in the system was 98%. In this case, duckweed pond system removed small amount of organic matter, because the UASB reactor was working practically as a secondary treatment. In the case of a real scale UASB reactor, higher organic loads will be transfer to the duckweed ponds but not as high as in the system without pretreatment, so no organic matter removal limitation will be expected.

Significant differences were found in the first three ponds when comparing same number ponds between the two systems. From pond 4 onwards the differences were not significant and further removal was very low. From above discussion it can be concluded that the system without pretreatment can be design with four ponds (HRT=12 days) to reach high efficiency of organic matter. In the case of the system with pretreatment, if the UASB reactor is very efficient the duckweed ponds will serve as a polishing step for remaining organics present in the effluent and its main function will be on the removal of nutrients.

Conductivity and solids
The conductivity of the UASB effluent (559 ± + 54 µS cm^{-1}) was higher than conductivity of un-treated artificial wastewater (293 ± 1.7 µS cm^{-1}). This may be explained by the hydrolysis of organic matter in the UASB reactor, generating short chain organic acids and other ions, which are soluble in water. The same effect was observed to happen in pond 1 of the system without pre-treatment.

In both systems the level of conductivity diminished gradually through the different ponds and produced an effluent of 432-434 µS cm^{-1}, this means a removal efficiency of 23%. As conductivity is an indirect measurement of dissolved solids and it is associated with the salinity of the water, a reduction in this parameter is important when the effluent is going to be re-used in irrigation.

In terms of suspended solids, the effluents of both systems were very good, usually under 10 mg l⁻¹. This is in agreement with the observed low level of algae. Other studies have also reported high quality duckweed ponds effluents in terms of suspended solids (Alaerts et al., 1996; Zimmo et al., 2002).

Nitrogen and Phosphorus
Nitrogen and phosphorus concentrations were adequate for duckweed growth in all ponds of both systems. The removal efficiencies increase gradually along the seven ponds in each system to reach 37 and 48 % for nitrogen and 45 and 50 % for phosphorus in the lines with and without pretreatment, respectively. The differences were not significant. These removal efficiencies are similar to those reported by Al-Nozaily (2001).

Complete hydrolysis and ammonification of the organic nitrogen occurred not only in the UASB reactor but in the first pond of the line without pretreatment. This is in disagreement with Alaerts et al. (1996), who reported that hydrolysis and ammonification of organic nitrogen was a limiting step for nitrogen removal in duckweed ponds. Nitrate was not detected in any of the systems during this experiment, even in the pond where oxygen concentrations were above 1 mg l⁻¹. It have been reported that oxygen concentrations higher than 1 mg l⁻¹ are needed to support an efficient nitrification process (Metcalf and Eddy, 1991). The absence of nitrate in the effluent could mean that nitrification was not present in the systems or that the rate of denitrification was higher or equal to the rate of nitrification. This indicates the importance of establishing a detailed nitrogen balance for the systems in order to have a better understanding of the nitrogen transformations happening in duckweed system. This information will lead to the generation of alternatives to improve nitrogen removal efficiency.

Conclusions
pH levels are very stable in both systems with and without anaerobic pretreatment. Temperature gradients are present during daytime but not as high as they may be in conventional stabilization ponds. Oxygen levels are significantly higher in the duckweed system with anaerobic pretreatment, especially in the top layer. Nevertheless, re-aeration rates are low in both systems.

Both systems were efficient in removing organic matter. The system without pretreatment obtained 98% of BOD₅ removal in pond 4, so 12 days of retention time will be enough to reach high organic matter removal. The system with pretreatment obtained also 98% BOD₅ removal (92% of BOD₅ in UASB reactor). In this case with a very efficient UASB reactor, the duckweed ponds will serve as a polishing step for remaining organic matter and its main function will be the removal of nutrients.

Nutrient removals were 37-48% for nitrogen and 45-50 % for phosphorus in the lines with and without pretreatment respectively. They were slightly higher in the system without pretreatment but no significant difference was found between the two systems. It is important to establish the nitrogen balance for the systems in order to generate a better understanding of the nitrogen transformations in the duckweed system.

Acknowledgements
The authors would like to thank Dr J. Sanabria, Engineering Faculty-Universidad del Valle for her valuable discussions and to J. Soto, A. Quintero and D. Timana, for helping to collect the data. The authors would like to acknowledge the Dutch Government and Universidad del Valle for their financial support via the cooperative project ESEE, which is funded by the SAIL programme.

References
Alaerts G., Mahbubar Rahman, Kelderman P. (1996). Performance Analysis of a full-scale duckweed-covered sewage lagoon. *Wat. Res.*, 30 (4), 843-852.

Al-Nozaily F. A. (2001). Pilot plant operation of a duckweed-covered sewage lagoon (DSL) in Sana'a, Yemen- II. Growth, nutrients budget and FC removal. . In: Performance and Process Analysis of Duckweed-Covered Sewage Lagoons for high Strength Sewage. Doctoral Dissertation. Delft University of Technology-International Institute of Hydraulic and Environmental Engineering. Delft-Holland.

A. P. H. A. (1995). American standard methods for the examination of water and wastewater. 19th edition. New York.

Bitcover E. H. and Sieling D. H. (1951). Effect of various factors on the utilisation of nitrogen and iron by *Spirodela polyrrhiza* (L.) Schleid. *Plant Physiol.* 26, 290-303.

Brix H. and Schierup H. H. (1989). The use of aquatic macrophytes in water pollution control. *Ambio*, 18, 100-107.

Cubillos A. (1994). Lagunas de Estabilización. Centro Interamericano de Desarrollo e Investigación Ambiental y Territorial (CIDIAT). Mérida, Venezuela. ISBN 980-221-085-6.

Gijzen H. and Khondker M. (1997). An overview of the ecology, physiology, cultivation, and application of Duckweed, Literature Review. Report of Duckweed Research Project. Dhaka, Bangladesh.

Gijzen H.J. and Ikramullah M. (1999). Pre-feasibility of duckweed-based wastewater treatment and resource recovery in Bangladesh. World Bank Report, Washington D. C.

Gijzen H. J. (2001). Anaerobes, aerobes, and phototrophs: a wining team for wastewater management. *Wat. Sci & Tech.*, 44(8), 123-132.

IMTA-Insttituto Mexicano de Tecnología del Agua, Comisión Nacional del Agua. (1984). Manual de diseño de agua potable, alcantarillado y saneamiento (Lagunas de Estabilización). México.

Landolt E., Kandeler R. (l987). The family of *Lemnaceae* monographic study, Vol.2. *Veroeffentlichungen des geobotanisches Institutes der ETH*, Stiftung Rubel, 95, Zurich, 1-638.

Mara D. D., Alabaster G. P., Pearson H. W., Mills S. W. (1992). Waste stabilization ponds. A design manual for Eastern Africa. Lagoon Technology International. Leeds, England.

Mendonca S. R. (1999). Lagunas de Estabilización. Organización Panamericana de la Salud (OPS-OMS). Bogotá, Colombia.
Metcalf and Eddy. (1991). Waste Engineering. Treatment, disposal and reuse. Tchobanoglous G. and Burton F. L. [eds.]. 2nd Ed. McGraw Hill, Inc. USA.

Moorhead K. K. and Reddy K. R. (1988). Oxygen transport through selected aquatic macrophytes. *Jour. Environ. Qual.*, 17, 138-142.

Oron G., Porath D., Wildschut L.R. (l986). Wastewater treatment and renovation by different duckweed species. *J. Env. Eng. Div., ASCE*, 112(2), 247-263.

Reed S. C., Middlebrooks E. J., Crites R. W. (1995). Natural systems for waste management and treatment. 2nd Ed. *Mc Graw Hill*. New York.

Skillicorn P., Spira W., Journey W. (l993). Duckweed aquaculture, a new aquatic farming system for developing countries. The World Bank. 76 p. Washington.

Zimmo O., Al-Sa'ed R. M., Van der Steen N. P, Gijzen H. J. (2002). Process Performance Assessment of algae-based and duckweed-based wastewater treatment systems. *Wat. Sci. Tech.*, 45(1), 91-101.

Zirscky J. and Reed S. (1988). The use of duckweed for wastewater treatment. *Jour. Wat. Poll. Cont. Fed.* 60 (7), 1253-1258.

Chapter 3

Effect of total ammonia nitrogen concentration and ph on growth rates of duckweed *(Spirodela polyrrhiza)*

Published as:

Caicedo J. R., Steen N. P. van der, Arce O., Gijzen H J. (2000). Effect of total ammonium nitrogen concentration and pH on growth rates of duckweed (*Spirodela polyrrhiza*). *Wat. Res.* 34(15), 3829-3835.

Chapter 3

Effect of total ammonia nitrogen concentration and ph on growth rates of duckweed (Spirodela polyrrhiza)

Abstract

The use of duckweed in domestic wastewater treatment is receiving growing attention in the last years. Duckweed based ponds in combination with anaerobic pre-treatment may be a feasible option for organic matter and nutrient removal. The main form of nitrogen in anaerobic effluent is ammonium. This is the preferred nitrogen source of duckweed but at certain levels it may become inhibitory to the plant. Renewal fed batch experiments at laboratory scale were performed to assess the effect of total ammonia ($NH_3 + NH_4^+$) nitrogen and pH on the growth rate of the duckweed *Spirodela polyrrhiza*. The experiments were performed at different total ammonia nitrogen concentrations, different pH ranges and in three different growth media. The inhibition of duckweed growth by ammonium was found to be due to a combined effect of ammonium ions (NH_4^+) and ammonia (NH_3), the importance of each one depending on the pH.

Key Words

Ammonia, ammonium, duckweed, growth inhibition, nitrogen, *Spirodela polyrrhiza*, toxicity, wastewater treatment.

Introduction

The use of aquatic macrophytes, such as water hyacinth, duckweed, water lettuce etc., in wastewater treatment has attracted global attention in recent years (Reed *et al.*, 1995; Gijzen and Khondker, 1997; Steen *et al.*, 1999; Vermaat and Hanif, 1998). These plants can be applied on the surface of stabilisation ponds, and may contribute to nutrient recovery from wastewater. Duckweed species have shown characteristics that make duckweed based systems (DBS) very attractive, not only for wastewater treatment but also for nutrient recovery. The reason for this is the rapid multiplication of duckweeds and the high protein content of its biomass (30-49% of dry weight; Oron *et al.*, 1984). Therefore duckweed can accumulate considerable amounts of nutrients that can be removed by simple and low cost harvesting technologies. The harvested duckweed may be used as a valuable fish or animal feed (Skillicorn *et al.*, 1993). Due to these characteristics, DBS have an important potential for resource recovery (Culley and Epps, 1973; Mbagwu and Adenihi, 1988).

Treatment efficiency of DBS for biological and chemical oxygen demand (BOD and COD) is similar to that of conventional stabilisation ponds (Bonomo *et al.*, 1997), but removal of suspended solids is usually better in DBS, due to suppression of algae growth (Van der Steen *et al.*, 1999). Another advantage of DBS is that nutrients are (partly) recovered rather then lost to the atmosphere or removed with the effluent. Area requirements of DBS to satisfy discharge standards for BOD may be reduced by pre-treatment in high rate anaerobic reactors. Anaerobic pre-treatment, for instance in an Up flow Anaerobic Sludge Blanket reactor, effectively reduces the BOD, but has negligible effect on bacterial pathogen counts and nutrient concentrations. Therefore anaerobic pre-treatment complemented with duckweed ponds may be a feasible low cost technology to achieve effective BOD and TSS removal and nutrient recovery (Alaerts *et al.*, 1996; Gijzen and Khondker, 1997).

Nitrogen compounds and duckweed growth

The nitrogen in anaerobic effluent is present mainly as ammonium (NH_4^+). This is an advantage because duckweed has a preferential uptake of ammonium over other sources of nitrogen (Porath and Pollock, 1982). However, the ammonium ions are inhibitory to duckweed growth at high concentrations (Oron *et al.*, 1984). The inhibition by total ammonia ($NH_4^+ + NH_3$) has commonly been attributed more to the NH_3 form than to the NH_4^+ form (Vines and Wedding, 1960; Warren, 1962). The pH of the growth medium or wastewater determines the ratio between the two species and therefore the NH_3 concentration. The un-dissociated and uncharged NH_3 molecule is lipid-soluble and therefore easily enters plant cells through their membrane and disturbs the cell metabolism. Biological membranes are relatively impermeable to the ionised and hydrated form NH_4^+ that is generally thought to be less detrimental for duckweed growth. However, Ingermarsson *et al.* (1987) suggested that high NH_4^+ concentrations result in strong depolarisation of the membrane. This could result in a general inhibition of anion transport.

Ammonia and ammonium inhibition of duckweed growth

The response of duckweed to ammonium and ammonia levels is reported extensively in literature, but the conclusions are not always in line with each other. Bitcover and Sieling (1951), using artificial growth medium, found toxicity effects on *Spirodela polyrrhiza* at concentrations above 46 mg N/l of total ammonia in the pH range 5 to 8. Rejmankova (1979, as cited by Wildschut, 1984) reported tolerance up to 375 mg/l of total ammonia nitrogen. Wildschut (1984) and Oron *et al.* (1984) found 200 mg/l of total NH_4^+-N in domestic wastewater as unfavourable to duckweed (*Lemna gibba*) growth, at pH 7. Wang (1991) studied the toxicity of the un-dissociated form (NH_3) on duckweed (*Lemna minor*) and a direct relationship was observed between un-dissociated ammonia concentration and the percentage of inhibition in renewal batch experiments with artificial substrate at initial pH = 8.5. An un-ionised ammonia concentration of 7.2 mg/l was calculated to cause 50% duckweed growth inhibition. It is difficult to compare the results of the studies mentioned above, since these were obtained under different conditions of temperature, pH, wastewater and medium composition and duckweed species.

The successful development and implementation of DBS for wastewater treatment depends among others on the duckweed yield. A high duckweed yield will result in the effective removal of nutrients from wastewater, while the application of duckweed as an animal feed could generate substantial revenues. In developing countries domestic wastewater often contains high concentrations of ammonium, due to low water consumption. This research therefore investigated the effect of ammonium, ammonia and pH on the growth of duckweed biomass in order to improve the design and operation of DBS for wastewater treatment and nutrient recovery.

Materials and methods

Experimental set up
The effects of total ammonia concentration, pH and type of growth medium on duckweed growth were studied under various experimental conditions (Table 1). These experiments were conducted in 250 ml plastic containers with a water depth of 5 cm that were operated as renewal fed batch reactors. At the beginning of each experiment ten healthy duckweed fronds from a stock culture were put in each of the containers. An experiment lasted for 14 days, during which the fronds were counted every 3 days. The total dry weight of the duckweed biomass was determined at the start and at the end of the experiments. The medium was replaced every 4 days to compensate nutrient losses and to reduce algae growth. In addition, the ammonium nitrogen levels were restored to the initial concentration every other day by adding NH_4Cl. The pH was measured every day and subsequently adjusted to the initial conditions with NaOH or HCL solutions. The average pH during a particular day was assumed to be the average of the pH measured just before the pH adjustment, and the pH that was set. The average pH for the total incubation period was calculated by taking the average of the daily average pH values, for the three duplicates. Average pH calculations were based on average H^+-concentrations. The pH range was defined as the range between the maximum and minimum daily average value. Containers were placed randomly under fluorescent lamps at a light intensity of 85-100 $\mu E\ m^{-2} \cdot s^{-1}$) (16 hours light, 8 hours dark). The average water temperature was 25 °C. Evaporation losses were compensated every day with tap water. Each treatment had three replicates.

Table 1. Experimental variables for the growth experiments with *Spirodela polyrrhiza*.

Growth medium	Initial pH	Initial Total Ammonia concentration $(mg\ N\text{-}NH_4^+ + N\text{-}NH_3 /l)$
10% Huttner medium	5, 7, 8, 9	3.5, 20, 50.100
UASB effluent	5, 7, 8, 9	20, 50, 100
Raw domestic wastewater	5, 7, 8, 9	20, 50, 100

Duckweed stock cultures
Duckweed (*Spirodela polyrrhiza*) was collected from a natural pond in the area of Cali, Colombia, and adapted to the different growth media in the laboratory of the Engineering Faculty of Valle University. Stock cultures for each growth medium were maintained in 1-liter containers and the growth medium was replaced every four days.

Growth media
Three different growth media were used in these experiments (see Table 1). Anaerobic effluent was obtained from a full scale Upflow Anaerobic Sludge Blanket

(UASB) reactor treating domestic wastewater. The original composition of Huttner media (Landolt and Kandeler, 1987) was modified such that ammonium nitrogen was the only source of nitrogen (Vermaat and Hanif, 1998): Ca $(NO_3).4H_2O$ was replaced by $CaCl_2$ and $CoNO_3$ by $CoSO_4$ (Table 2). Raw domestic wastewater and the UASB effluent samples were obtained from the inlet to the sewage works in Cali and from a full scale UASB reactor on the same location, respectively. Composite samples were taken once a week during 8 hours (Table 3).

Table 2. The composition of 10% modified Huttner medium (Vermaat and Hanif, 1998).

Nutrient compound	Concentration (mg/l)
C_aCl_2	12.2
EDTA (di-sodium salt)	50.0
K_2HPO_4	40.0
NH_4NO_3	20.0
$ZnSO_4.7H_2O$	6.5
H_3BO_3	1.5
$Na_2MoO_4.H_2O$	2.5
$CuSO_4.5H_2O$	0.4
$CoSO_4.7H_2O$	0.02
$MnCl_2.4H_2O$	3.5
$FeSO_4.7H_2O$	2.5
$MgSO_4.7H_2O$	50

Table 3. The composition of domestic wastewater and UASB effluent*

Parameters	Domestic wastewater	UASB effluent
Chemical oxygen demand, COD (mg/l)	270-648	104-169
Biochemical oxygen demand, BOD_5 (mg/l)	121-263	37-78
Total Kjeldahl nitrogen (mg N/l)	29-60	23-32
NH_4^+-N (mg N/l)	22-32	17-29
PO43--P (mg P/l)	0.3-1.6	0.16-1.44

* source: Wastewater control section of the Municipal Authority of Cali.

Physico-chemical analysis
Total ammonia concentrations (NH_4^+-N + NH_3-N) and pH were determined with ion selective electrodes. The average values for these parameters during each experiment were used to calculate the characteristic NH_3 concentration, by using the equations 1, 2, and 3:

$$K_a = \frac{[H^+][NH_3]}{[NH_4^+]} \qquad (1)$$

Where K_a is the ammonia dissociation constant. The concentration of each species is also temperature dependent as described by the following equation (Emerson *et al.*, 1975):

$$pK_a = 0.09108 + \frac{2729.92}{273.2 + T} \qquad (2)$$

Where T is temperature in °C. By modifying equation (1), the fraction NH_3 of the total ammonia concentration can be expressed as follows (Pano and Middlebrooks, 1982):

$$\%Un-ionized\ NH_3 = \frac{100}{1+10^{(pKa-pH)}} \qquad (3)$$

Growth analysis
Duckweed growth was evaluated on the basis of the relative growth rate (RGR) as given by:

$$\ln(N_t) = \ln(N_o) + RGR * t \qquad (4)$$

Where,

N_t= number of fronds or dry weight, at time t.
N_o= number of fronds or dry weight, at time 0.

The RGR was determined in two ways: Firstly on the basis of the frond numbers at the beginning and end of the experiments, and secondly on the basis of the dry weight biomass at the beginning and end of the experiments. The RGR was calculated by fitting equation (4) to those values. The fronds were actually counted every 3 days and growth appeared exponential during the duration of the experiment and therefore equation (4) could be applied. Initial dry weight was determined by drying five representative samples of 10 healthy fronds at the beginning of the experiments. Dry weight of the plants was measured after drying the plants at 70 °C until constant weight (Vermaat and Hanif, 1998).

Results
The RGR of *Spirodela polyrrhiza* under the experimental conditions was found to decrease with increasing concentrations of total ammonia (Fig. 1). The RGR based on frond counting was similar to the RGR based on dry biomass production only for the lowest ammonium concentration (3.5 mg/l N). The decrease of the RGR at higher ammonium concentrations was stronger for the RGR based on dry biomass production. This is illustrated in Figure 1 for Huttner medium at an average pH of 7.3, but similar differences were found in the other incubations.

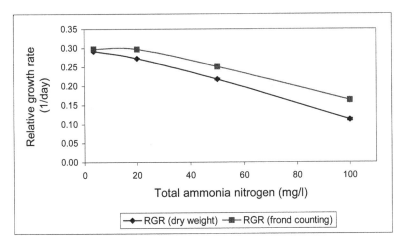

Fig. 1. Comparison of RGR based on frond number counting (○) and RGR based on dry weight biomass production (●) for incubation in Huttner media with pH 7-8. Note: the standard errors are smaller then 0.004 and therefore not visible in the graph.

The difference in RGR is explained by changes in the size and specific dry weight of duckweed fronds, as a result of total ammonia concentration. This effect seems to be more pronounced at higher pH values (Fig. 2). Since RGR expressed on the basis of dry weight is a more direct indication of growth, this unit was used in all further experiments. The effect of various combinations of average pH and total ammonia concentration on the RGR is shown in Figure 3. The pH fluctuated considerably during the incubations, even with daily pH adjustments (Table 4).

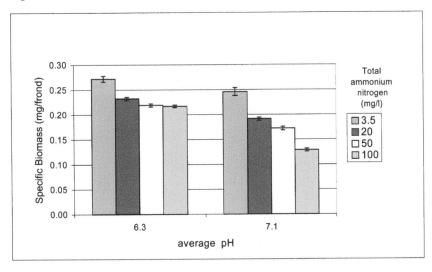

Fig 2. Effect of total ammonium nitrogen concentration on the specific frond weight of Spirodela polyrhiza. Standard errors are indicated.

The duckweed was able to grow at total ammonia concentration above 50 mg/l N, but only if the average pH was lower than 7.9 for Huttner medium and lower than 8.5 for the other media. Duckweed died at 50 and 100 mg/l ammonium nitrogen when the pH fluctuated between 7.4 and 9 in all three media. A general trend for the incubations with average pH 5.9-7.4 is that the RGR is decreasing with both increasing pH and increasing total ammonia concentrations.

Table 4. The average pH values and the ranges observed in the incubations.

Initial	10 % Huttner medium		UASB effluent		Domestic wastewater	
pH	Average	Range	Average	Range	Average	Range
5	4.2	3.1-5.1	4.8	4.4-5.1	4.7	4.1-5.3
7	5.9	5.0-7.0	6.6	5.9-7.2	5.7	4.9-7.2
8	7.3	6.9-8.0	7.3	6.5-8.0	7.4	6.6-8.0
9	7.9	7.4-9.0	8.5	8.3-9.0	8.5	8.0-9.1

Concentration of un-dissociated ammonia

The average NH_3 concentrations in the incubations were calculated on the basis of the initial total ammonia concentration and the average pH value, according to eq. 1, 2 and 3. The relation between RGR and NH_3 concentration for the incubations with average pH values between 5.9 and 7.4 is shown in Figure 4. In this pH range duckweed has been reported to grow well (Landolt and Kandeler, 1987), whereas extreme pH values may cause direct growth inhibition, independent from ammonium effects. Some incubations with high NH_4^+ concentrations and low NH_3 concentrations (due to relatively low pH values) showed lower RGR than incubations with low concentrations of both NH_3 and NH_4^+ (Fig. 4). Assuming no direct pH effects on the RGR in the incubation with average pH 5.9-7.4, it seems that not only NH_3 but also NH_4^+ negatively affects the RGR.

Therefore a multiple linear regression was done for the incubations in the three different media, with average pH 5.9-7.4. The data from the incubations with pH values that fluctuated below 5 and above 8 were therefore not included in the regression. The regression equation was as follows:

$$RGR = a - b \, [N\text{-}NH4+ + N\text{-}NH3] - c \, [N\text{-}NH3]. \qquad (5)$$

Which equals,

$$RGR = a - b \, [N\text{-}NH_4^+ + N\text{-}NH_3] - c \, \{ \, [N\text{-}NH_4^+ + N\text{-}NH_3] \,.(100/(1+10^{pKa-pH})) \}. \qquad (6)$$

Where a, b and c are constants. As the regression equations found for the three media are quite similar an overall correlation for the combined data of the three media was also performed (Table 5). For Huttner medium and UASB effluent the constants for ammonium and ammonia were not significant, but for domestic wastewater and for the combined data the constants for ammonium and ammonia both were significant. The linear model for the combined data could explain 48% of the observed variation (R^2-adjusted = 0.48).

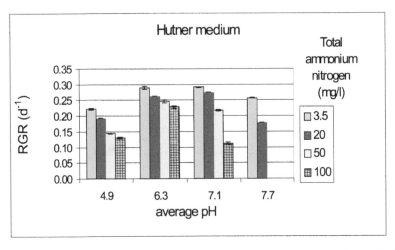

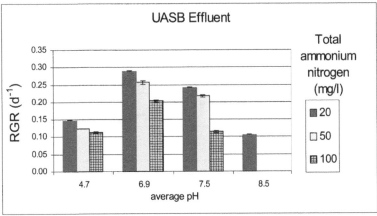

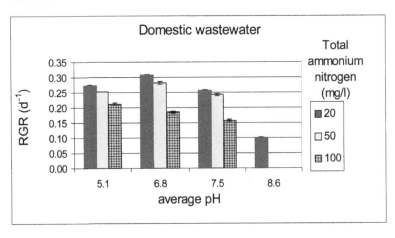

Fig 3 Effect of total ammonia on relative growth rates (based on dry weight measurements) in three different media at four pH ranges. The missing bars at average pH values of 7.9 and 8.5 indicate duckweed die-off. Standard errors are indicated.

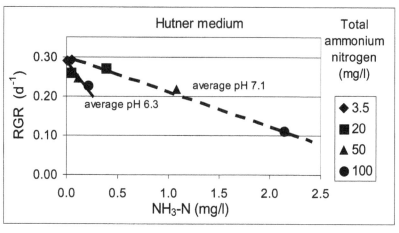

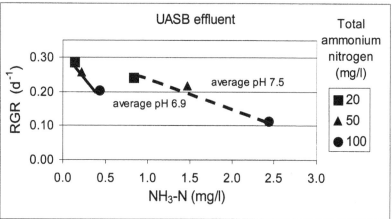

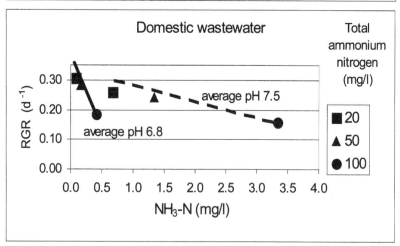

Fig. 4. Effect of NH$_3$ concentration on the RGR based on dry weight measurements.

Table 5. The results for the multiple linear regression according to equation 5

Medium	Regression constants and significance:			Adjusted R^2	Significance of model
	a	b	c		
Huttner medium	0.27 $p < 0.001$	0.00095 $p = 0.014$	0.020 $p = 0.168$	0.46	$p = 0.014$
UASB effluent	0.24 $p = 0.001$	0.00075 $p = 0.251$	0.021 $p = 0.156$	0.16	$p = 0.225$
Domestic wastewater	0.30 $p < 0.001$	0.00082 $p = 0.011$	0.033 $p < 0.001$	0.85	$p = 0.001$
Combined data	0.27 $p < 0.001$	0.00081 $p = 0.001$	0.025 $p < 0.001$	0.48	$p < 0.001$

Discussion

RGR determination methods
Two methods are reported in the literature for measurement of RGR, i.e. based on frond number counting and based on dry weight measurements (Clement and Bouvet, 1993; Wang, 1991; Körner and Vermaat, 1998). The present experiments showed that the growth inhibition by increasing ammonium concentrations was more pronounced for the RGR based on dry weight measurements (Fig. 1). Probably this is because not only the rate of frond reproduction was affected by the increase in ammonium concentrations, but also the size and specific weight of the fronds (Fig. 2). The reduction in specific weight is reflected in the RGR values based on dry weight biomass production but not in the RGR calculated on the basis of frond number counting. Therefore it is suggested that the RGR based on dry weight biomass production is the preferred parameter to assess the effect of ammonium on duckweed growth.

Combined effect of NH_3, NH_4^+ and pH
To assess the separate effect of the NH_3, NH_4^+ and pH on duckweed growth is difficult because the three parameters are interrelated by a chemical equilibrium. The pH determines the ratio between the NH_3 and NH_4^+ concentrations and therefore the presumed growth inhibition by these compounds. However, the pH fluctuations in some incubations reached values that may be directly detrimental for duckweed growth (<5 and >8). In new investigations in our laboratories this is prevented by addition of buffers. At pH <6, biomass production was observed in all incubations (Fig 3), but the fronds appeared unhealthy, wrinkled and yellowish. At pH values above 8 even duckweed die-off was observed in several incubations. The optimal pH value reported in literature for growth of *Spirodela polyrrhiza* is around 7 (Bitcover and Sieling, 1951; Landolt and Kandeler, 1987) and this was confirmed by the present experiments.

The maximum growth rates (around 0.3 d^{-1}) were observed in all three media for the lowest total ammonia concentrations (3.5 - 20 mg/l N) and are comparable with those reported in literature (Oron *et al.*, 1987; Vermaat and Hanif, 1998). The three media apparently contain enough nutrients (for instance P-PO_4) to sustain in principle maximum growth. Although P-PO_4 was much lower in UASB effluent and domestic wastewater than in Huttner medium, the same maximum growth rate was observed.

The incubations with pH values in the range where the pH is assumed not to affect directly the duckweed growth (6-8) showed that growth inhibition increased with increasing total ammonia concentrations and also with increasing pH values. Several researchers have explained this phenomenon as a result of the simultaneous increase in NH_3 concentration (Azov and Goldman, 1982; Wang, 1991). However, some incubations with relatively high NH_3 concentrations showed higher growth rates as compared to incubations with lower NH_3 concentrations (Fig. 4). This could be explained by assuming that the relatively high concentrations of NH_4^+ in the latter incubations negatively affected duckweed growth. Apparently both forms of ammonium cause growth inhibition in duckweed. The regression coefficients (Table 5) indicate that NH_3 inhibition occurs at much lower concentrations than NH_4^+ inhibition. In practice this is compensated by the fact that NH_3 is only a small fraction of the total ammonia.

Both inhibition mechanisms have been reported in literature. The NH_3 inhibition stems from its inherent toxicity and its potential to easily cross cell membranes (Vines and Wedding, 1960; Warren, 1962). The growth inhibition by NH_4^+ is due to saturation and depolarisation of cell membranes, thus inhibiting anion transport (Ingermarsson et al., 1987). It was not always taken into account by other researchers that the inhibition of total ammonia on duckweed growth is due to both ammonium forms, and moreover that the ratio of those species is strongly pH dependent. This may explain the diversity of results found in literature with respect to ammonium inhibition (Wildschut, 1984; Wang, 1991).

Consequences for duckweed ponds in practice
The application of duckweed ponds as a sustainable technology for wastewater treatment and resource recovery has received increasing attention over the last years, but not much is known yet about how to design ponds in order to achieve maximum duckweed production and nutrient conversion. Our study has shown that ammonium concentrations and pH values that are commonly found in domestic wastewater may severely reduce duckweed growth. If growth inhibition should be controlled at or below 30% then, total ammonia concentrations in the duckweed pond should be below 50 mg/l, while pH should be maintained below 8. For ammonium concentrations in pond water between 50 and 100 mg/l N, it seems that the pH should not be higher than 7.

The design of duckweed pond systems should be adjusted to the influent ammonium concentration, pH value and buffering capacity. A plug flow pond system is not recommendable if the influent is expected to cause inhibition. The first pond of a duckweed system should rather be well mixed and designed in such a way that the optimum NH_4^+ concentration is maintained. Recycling of the effluent into the first pond may also be applied to maintain optimum growth conditions and to increase system stability. This research indicated that for batch conditions this optimum NH_4^+ concentration is below 20 mg/l N. Presently research is conducted in our laboratories to investigate if the same applies to continuous flow systems.

Conclusions

◆ The effect of parameters such as total ammonia concentration and pH on duckweed growth should preferably be assessed by measuring the RGR based on dry biomass production, rather than the RGR based on frond number counting.

◆ The maximum RGR was observed at low concentrations of ammonium (3.5 - 20 mg/l N). In the pH range where no direct effects of pH are expected (5-8), it was found that both increasing total ammonia concentration and increasing pH values caused increasing growth inhibition.

◆ The inhibition of duckweed growth by total ammonia nitrogen is probably the result of a combined effect of two inhibition mechanisms, due to ammonium and ammonia, respectively. The relative importance of each mechanism is determined by the pH.

Acknowledgements

This research was developed at Universidad del Valle, Cali, Colombia in the context of the cooperative project IHE/DUT/Univalle, project no. C00003303, which received financial support from SAIL, via The Netherlands Development Agency (NEDA), Ministry of Foreign Affairs, The Netherlands. The authors thank prof. Guy Alaerts and ir. Siemen Veenstra for their contributions to the development of this research line and Dr. Jan Vermaat for critically reading the manuscript.

References

Alaerts, G., Mahbubar, R. and Kelderman, P. (1996) Performance Analysis of a full-scale duckweed-covered sewage lagoon. *Wat. Res.* 30 (4), 843-852.

Azov, Y. and Goldman, J.C. (1982) Free ammonia inhibition of algal photosynthesis in intensive cultures. *Appl. Environ. Microbiol.* 43 (4), 735-739.

Bitcover, E.H. and Sieling, D.H. (1951) Effect of various factors on the utilisation of nitrogen and iron by *Spirodela polyrrhiza*. *Plant Physiol.* 26, 290-303.

Bonomo, L., Pastorelli, G. and Zambon, N. (1997) Advantages and limitations of duckweed-based wastewater treatment systems. *Wat.Sci.Tech.* 35 (5), 239-246.

Clement, B. and Bouvet, Y. (1993) Assessment of landfill leachate toxicity using the duckweed *Lemna minor*. *The Science of the Total Environment*, Supplement, 1179-1190.

Culley, D.D. and Epps, E.A. (1973) Use of duckweed for waste treatment and animal feed. *Jour. Wat. Poll. Cont. Fed.* 45 (2), 337-37.

Emerson, K., Russo, R.E., Lund, R.E. and Thurston, R.V. (1975) Aqueous ammonia equilibrium calculations: Effect of pH and temperature. *Jour.Fisheries Res. Board of Canada* 32 (12), 2379-2383.

Gijzen, H. and Khondker, M. (1997) An overview of the ecology, physiology, cultivation, and application of Duckweed, Literature Review. Report of Duckweed Research Project. Dhaka, Bangladesh.

Ingermarsson, B., Oscarsson, P., Ugglas, M.A. and Larsson, C. M. (1987) Nitrogen utilisation in *Lemna*, III. Short term effects of Ammonium on nitrate uptake and nitrate reduction. *Plant Physiol.* 85, 865-867.

Körner, S. and Vermaat, J. (1998) The relative importance of *Lemna gibba* L., bacteria and algae for the nitrogen and phosphorus removal in duckweed-covered domestic sewage. *Wat.Res.* 32 (12), 3651-3661.

Körner , S, Das, S.K., Vermaat J.E. and Veenstra S. (in prep.) Ammonia toxicity to the duckweed *Lemna gibba* used for the treatment of wastewater.

Landolt, E. and Kandeler R. (1987) The family of *Lemnaceae* monographic study, Vol.2. *Veroeffentlichungen des geobotanisches Institutes der ETH*, Stiftung Rubel, 95, Zurich, 1-638.

Mbagwu, I.G. and Adeniji, H.A. (1988) The nutritional content of duckweed (*Lemna puntata*) in the Kainji Lake area, Nigeria. *Aquat. Bot.* 29, 357-366.

Oron, G., Wildschut, L.R. and Porath, D. (1984) Waste water recycling by duckweed for protein production and effluent renovation. *Wat.Sci.Tech.* 17, 803-817.

Oron, G., Porath, D., Jansen, H. (1987) Performance of the duckweed species *Lemna gibba* on municipal wastewater for effluent renovation and protein production. *Biotech.& Bioeng.* 29 (2), 258-268.

Pano, A. and Middlebrooks, E.J. (1982) Ammonia nitrogen removal in facultative wastewater stabilisation ponds. *Jour. Wat.Poll. Cont. Fed.* 54 (4), 344-351.

Porath, D. and Pollock, J. (1982) Ammonia stripping by duckweed and its feasibility in circulating aquaculture. *Aquat.Bot.* 13, 125-131.

Reed, S.C., Crites, R.W. and Middlebrooks, E.J. (1995) Natural systems for waste management and treatment, Second edition. McGraw-Hill, Inc., New York.

Skillicorn, P., Spira, W. and Journey, W. (1993) Duckweed aquaculture. World Bank publication, Washington.

Steen, N. P. van der, Brenner, A., Van Buuren, J. and Oron, G. (1999) Post-treatment of UASB reactor effluent in an integrated duckweed and stabilization pond system. *Wat. Res.* 33 (3), 615-620.

Vermaat, J.E. and Hanif, K.M. (1998) Performance of common duckweed species (*Lemnaceae*) and the waterfern *Azolla filiculoides* on different types of waste water. *Wat. Res.* 32 (9), 2569-2576.

Vines, H.M. and Wedding R.T. (1960) Some effects of ammonia on plant metabolism and a possible mechanism for ammonia toxicity. *Plant Physiol.* 35, 820-825.

Wang, W. (1991) Ammonia toxicity to macrophytes (common duckweed and rice) using static and renewal methods. *Environ. Tox. Chem.* 10, 1173-1177.

Warren, K.S. (1962) Ammonia toxicity and pH. *Nature* 195, 47-49.

Wildschut, L.R. (1984) Introductory study on the performance of *Lemnaceae* on fresh municipal wastewater with emphasis on the growth of the *Lemnaceae*, the ammonium removal from the wastewater. *Doktoraal verslagen serie 84-3. Wageningen University.*

Chapter 4

The effect of anaerobic pre-treatment on the performance of duckweed stabilization ponds.

Published as:

Caicedo J. R., Steen N. P., Gijzen H. J. (2003). The effect of anaerobic pre-treatment on the performance of duckweed stabilization ponds. Proceedings of International Seminar on natural systems for wastewater treatment. Agua 2003. Cartagena Colombia.

Chapter 4

The effect of anaerobic pre-treatment on the performance of duckweed stabilization ponds.

Abstract
The effect of anaerobic pre-treatment on the performance of a duckweed stabilization pond system was assed in a pilot plant located in the Ginebra Research Station-Colombia. The pilot plant consisted of two lines of seven duckweed ponds in series. One line received de-gritted domestic wastewater and the other received effluent of a UASB reactor. Both lines were operated at a total hydraulic retention time of 21 days. The systems were monitored for temperature, pH, oxygen, biochemical oxygen demand, chemical oxygen demand, total suspended solids, total phosphorus, and different forms of nitrogen and biomass production. No effect of anaerobic pretreatment was observed on pH and temperature in the two systems. Oxygen concentrations were higher in the system with UASB reactor. Although both systems complied with the Colombian regulation in terms of organic matter (> 85% removal), the pretreatment with UASB reactor may contribute to the reduction of area requirement. Effluent quality in terms of total suspended solids was excellent, 9 ± 2 and 4 ± 1 mg l^{-1} in system with and without pre-treatment, respectively. Total nitrogen removals were 63 % and 68% and phosphorus removals were 24% and 29% in the system with and without pre-treatment, respectively. The differences were found no significantly different. Biomass production was in the range of 54-90 g m^{-2}-d^{-1} (fresh weight) in the system with pre-treatment and 36-84 g m^{-2}-d^{-1} in the system without pre-treatment. Total biomass productions were significantly different at 92% level of confidence. Protein content was 35.1% and 36.6% for the system with and without pre-treatment, respectively. Further research will be focused on understanding nitrogen transformations and removal mechanisms in duckweed covered sewage stabilization ponds.

Key words.
Duckweed ponds, *Lemnaceae*, *Spirodela polyrrhiza*, stabilization ponds, domestic wastewater, treatment efficiency, anaerobic pre-treatment.

Introduction
The need to develop sustainable technologies for domestic wastewater treatment is increasing rapidly, as the population grows and water supply coverage and consumption increases especially in developing countries. This need for sustainable technologies is especially acute in these countries, where the lack of economical resources makes it very difficult to implement expensive conventional treatment technology.

Waste stabilization ponds (such as algae, water hyacinth or duckweed ponds) have received increased attention in recent years as potential routes for making wastewater renovation more sustainable. Duckweed has excellent qualities like high protein content, high growth rate and is easy to handle. The possibility of simultaneously treating wastewater and producing duckweed in a pond system is an attractive option to contribute to alleviate food and water shortage.

The feasibility of the duckweed based pond system for domestic wastewater treatment has been documented in the literature (Reddy and DeBusk, 1985; Skillicorn et al., 1993; Alaerts et al., 1996; Al Nozaily, 2001; Zimmo et al., 2002). It has been applied at full-scale in Taiwan, China, Bangladesh, Belgium and the USA (Edwards et al., 1992; Zirschky and Reed, 1988; Alaerts et al., 1996). In many countries it is regarded as a polishing step for effluents of conventional treatment systems (Metcalf and Eddy, 1991).

In Colombia, duckweed based pond technology has been tested at pilot scale with artificial domestic wastewater (Caicedo et al., 2000). The results have shown that this technology is a feasible alternative for conventional wastewater treatment. The interest of the current work was to study the effect of anaerobic pre-treatment on the performance of a duckweed pond system treating domestic wastewater in order to define if the combination UASB reactor and duckweed pond has advantages over a system without anaerobic pre-treatment.

Materials and methods

Experimental set up
The pilot plant is located in the Wastewater Treatment station of Ginebra, a small municipality located in southwest of Colombia with about 8.000 inhabitants in the urban area. The village has a tropical climate with an average temperature of 24-28 °C. The wastewater is collected and transported from the town to the plant. The Ginebra Station is an experimental research and demonstration facility where different treatment technologies are being investigated like Conventional Stabilization Ponds, Up-flow Anaerobic Sludge Blanket Reactor (UASB), Anaerobic Filter, Wetlands, and Aquatic Macrophytes Ponds.

The pilot plant consisted of two lines of seven duckweed ponds in series. One line received raw domestic wastewater that had been de-gritted. The other line received effluent of the UASB Reactor (Fig. 1) The UASB Reactor had a hydraulic retention time of 7-8 hours. Each duckweed pond was made out of a plastic cylinder tank of 90-cm. height and 57.5-cm. diameter.

Operational Methods
In the Duckweed system, the water level of each tank was kept at 70 cm height. The hydraulic retention time (HRT) was 3 days for each tank, with a total HRT of 21 days for each line. The ponds were covered with duckweed of the specie *Spirodela polyrrhiza*. The domestic wastewater composition shows hourly changes throughout the day. To determine the average composition of Ginebra wastewater, it was necessary to conduct 24-hour sampling programs. The day was divided in periods of four or six hours. During each period a composite sample was collected by taking every half an hour a fixed-volume of sample and added all together for analysis. The

fluctuations in composition of the UASB effluent are less pronounced than for the wastewater but are still significant. Therefore, the same sampling program was also performed to determine its composition. Given the long retention time in the duckweed systems, enough to compensate the fluctuations of the influent, the effluent samples were grab samples taken from the effluent of each pond. Organic matter (BOD$_5$ and COD) and suspended solids were analyzed every two weeks and nitrogen and phosphorus compounds every week. The experiment lasted for three months. The influent and effluent flows were compared during several days and less than 1% difference was found on daily basis.

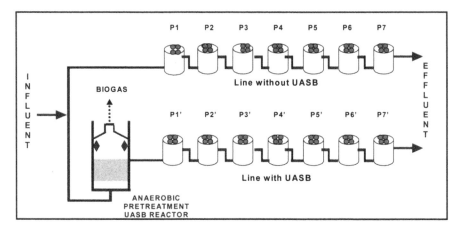

Fig. 1. Diagram of pilot plant. One line received domestic wastewater and the other line received the effluent of a UASB reactor.

Part of the duckweed cover was harvested every four days. Biomass samples were taken with a strainer from a known area and it was allowed to drain for about 5 minutes. The fresh weight was determined and the density was calculated. Enough biomass was harvested to leave a density of 700 g m^{-2} (fresh weight) after each harvesting. This density produced a full cover of the water surface and helped in preventing algae growth.

The data were compared and analyzed statistically; results for different treatments lines were compared using a parametric method (ANOVA test) which assumed independence of the compared values and a nonparametric method (Krush and Wallis test) (Daniel, 1990) which does not assumed independence. The results with these two tests were very similar. The results within each line were considered not independent and were compared with a paired t-Student test.

Analytical Methods
Chemical Oxygen Demand (COD), 5 days Biochemical Oxygen Demand (BOD$_5$), Alkalinity, Total Suspended Solids (TSS), Total Kjeldahl Nitrogen (TKN), Ammonium Nitrogen (NH$_4^+$-N), Nitrate Nitrogen (NO$_3^-$-N) and Nitrite Nitrogen (NO$_2^-$-N) were measured according to the Standards Methods (APHA, 1995) to no-

filtered samples. Dissolved Oxygen, pH, Temperature and Conductivity were measured with electrodes. Fecal coliform were determined using the membrane filtration method with Cromocult medium from Merck as the growth medium.

Results

Wastewater Characteristics
The average results of the 24-hours composite sampling programs to characterize the raw wastewater and the effluent of the UASB reactor are presented in Table 2.

Table 2 Characteristics of the Raw Wastewater and UASB Effluent.

PARAMETER	UNITS	RAW WASTEWATER*	UASB EFFLUENT*
pH	-	7.04 ± 0.15	6.75 ± 0.16
Temperature	°C	24.6 ± 1.2	24.6 ± 1.3
COD	mg l^{-1}	381 ± 99	189 ± 49
BOD$_5$	mg l^{-1}	254 ± 55	111 ± 23
TKN	mg l^{-1}	38.2 ± 4.7	36.8 ± 5.5
N-Ammonium	mg l^{-1}	26.4 ± 4.4	30.5 ± 5.3
N-Organic	mg l^{-1}	11.9 ± 4.4	6.3 ± 2.2
N-Nitrites	mg l^{-1}	0.04 ± 0.02	0.02 ± 0.008
N-Nitrates	mg l^{-1}	0.04 ± 0.02	0.02 ± 0.011
Total Phosphorus	mg l^{-1}	6.8 ± 1.8	6.8 ± 2.0
Total Solids	mg l^{-1}	458 ± 61	359 ± 24
Suspended Solids	m g l^{-1}	157 ± 46	57 ± 15
Conductivity	µS cm^{-1}	537 ± 51	624 ± 31

* Average ± S.D. (n = 12 of 24 hr integrated data; each integrated data = average of 4 or 6 sampling periods per day)

Temperature, pH and Oxygen Profiles
In previous experiments it was observed in duckweed ponds that temperature, pH and oxygen are more stable than in conventional stabilization ponds (Caicedo *et al.*, 2001). The vertical profiles (Depths: 7 cm, 35 cm, and 63 cm) and longitudinal profiles (along the seven ponds) were determined in the morning (8:30 h) and in the afternoon (16:30 h).

Temperature
In the morning, the temperature levels were in the range of 22- 23 °C. Along each system there were not significant differences between ponds. Within each pond,

although there was a significant difference between the upper and middle layer the vertical temperature gradients were very small, usually less than 1 °C. In the afternoon the water warmed up, especially in the top layers and there were significant differences between top, middle and bottom layers. The temperature ranged from 27 °C at the surface to 24 °C close to the bottom.

pH
The pH did not show significant vertical stratification (but the pond 1 in system with pre-treatment), either in the morning or in the afternoon. The pH along the system ranged between 6.6 and 7.6. It was close to neutral at the beginning of the system, showed a tendency to increase towards the middle and a tendency to decrease towards the last pond. Figure 2 presents average pH values for morning and afternoon measurements. The small variation during the day confirmed the low concentration of algae in these ponds. These results are quite different compared to conventional (algae) stabilization ponds, where it is very common to find variations of several pH units during the day time.

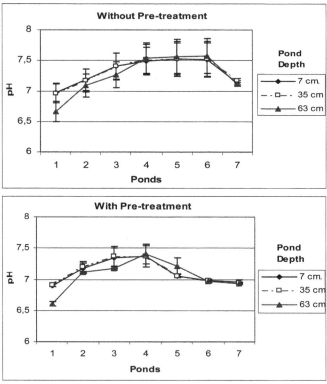

Fig. 2. pH profiles in duckweed pond systems fed with de-gritted or with anaerobically pre-treated domestic wastewater (n = 7). Standard errors are indicated.

Oxygen

The level of oxygen behaved very similar in both systems for the first four ponds. In these ponds, significant differences were found between top and middle layer in the same pond, and not significant differences were found between ponds of the same number pond in each line at the same depth. After pond 4 oxygen concentrations increased considerably, there were significant differences between systems (same pond number). Within each pond the differences were significant between top and middle layer for the system without pre-treatment and between top, middle and bottom layer in the system with pre-treatment. It can be observed in Figure 3 that the aerobic zone increased towards the bottom in the system with pre-treatment.

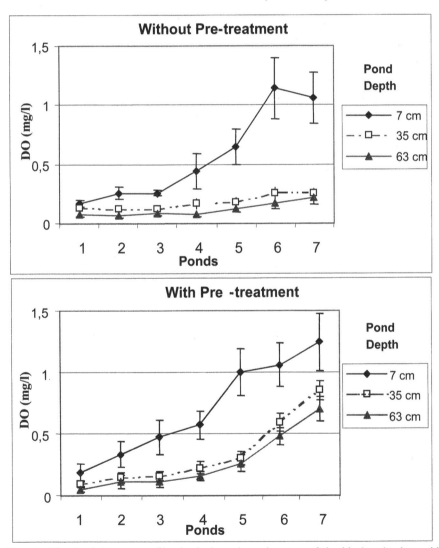

Fig. 3. Dissolved Oxygen profiles in duckweed pond systems fed with de-gritted or with anaerobically pre-treated domestic wastewater (n = 7). Standard errors are indicated.

Organic matter and suspended solids removal

In the system without pre-treatment organic matter (BOD_5, COD) was progressively removed along the system up to pond 5 (HRT= 15 d) (Fig. 4). The highest removal was present in the first pond (59% BOD_5, 55% COD). Not significant differences were found between the last three ponds in this system. It removed on average 90% of COD and 93% of BOD_5, producing an effluent with COD of 38±6 mg l^{-1} and BOD_5 of 18±4 mg l^{-1}

In the system with pre-treatment, the highest removal was on the UASB reactor (56 % BOD_5, 50 % COD), followed by the first pond (23% BOD_5, 27% COD). It was observed a significant removal until pond 5 (HRT= 15 d). A significant increase of COD was observed between ponds 6 and 7. This may be due to the smaller size of the duckweed fronds in pond 7 which allowed more light to enter into the very clear water column, where an increase of algae growth was observed. This reflected in COD and BOD_5 results, as the samples were not filtered. This system reached similar effluent quality as that of the system without pre-treatment with 89% COD removal, 88% BOD_5 removal and with effluent COD concentrations of 42±8 mg l^{-1} and BOD_5 of 29±6 mg l^{-1} .

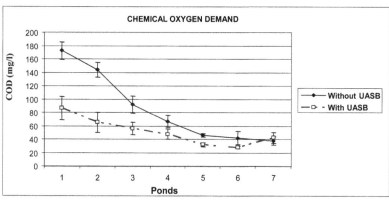

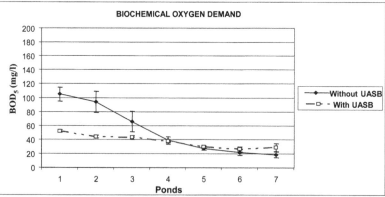

Fig. 4. Removal of COD (a) and BOD (b) in duckweed pond systems fed with de-gritted or with anaerobically pre-treated domestic wastewater. Standard errors are indicated.

The total suspended solids concentrations were under 15 mg l^{-1} along the system with the exception of the first pond in both systems. The final effluent concentrations were 9 ± 2 mg TSS l^{-1} (96% removal) and 4 ± 1 mg TSS l^{-1} (97% removal) in the systems with and without pre-treatment respectively.

Nitrogen and Phosphorus removal
Nitrogen concentrations in the effluent, in its different forms were monitored along each system. The results are presented in Figure 5 and 6.

Effluent ammonium nitrogen was significantly lower in the system with pretreatment (5.9 ± 2.1 mg NH_4^+-N l^{-1}, 78 % removal) than in the system without pre-treatment (8.5 ± 1.0 mg NH_4^+-N l^{-1}, 68% removal). Effluent Kjeldahl nitrogen was not significantly different between the two systems (10.5 ± 3.0 mg TKN l^{-1}, 71% removal for system with pre-treatment; 10.8 ± 1.3 mg TKN l^{-1}, 70% removal for system without pre-treatment).

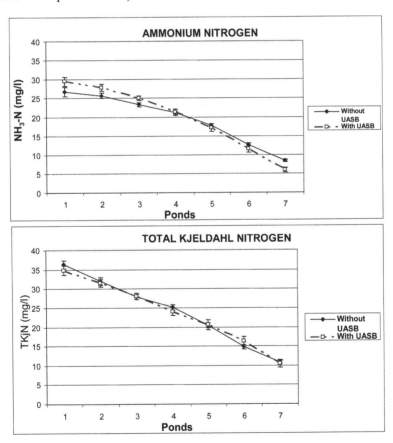

Fig. 5. Effluent reduced nitrogen concentrations in duckweed pond systems fed with de-gritted or with anaerobically pre-treated domestic wastewater (n = 10). Standard errors are indicated.

Oxidized nitrogen was very low in the first three ponds of both systems and started to increase from pond 4 (Fig. 6). For nitrates, there were significant differences between systems (same pond number) in the last four ponds. Effluent nitrate concentrations were 3.0±-0.1 mg NO_3-N l^{-1} and 1.3±0.1 mg NO_3-N l^{-1} for systems with and without pre-treatment, respectively.

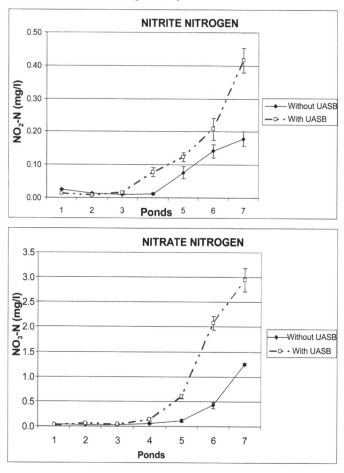

Fig. 6. Effluent oxidized nitrogen concentrations in duckweed pond systems fed with de-gritted or with anaerobically pre-treated domestic wastewater (n =10). Standard errors are indicated.

Effluent total nitrogen concentrations were 13.8 ± 2.9 mg N l^{-1} (64% removal) and 12.2 ± 1.2 mg N l^{-1} (68% removal), in the systems with and without anaerobic pre-treatment, respectively. Statistical analysis showed no significant difference between the two systems. Within each system, nitrogen removals were significant in every pond and the rates were fairly constant along the system.

Phosphorus concentration gradually decreased along both systems (Fig. 7). The removal efficiencies were 24% and 29% and the effluent concentrations were 5.2 ± 0.6 mg P l^{-1} and 4.8 ± 0.6 mg P l^{-1} in the systems with and without pretreatment, respectively. Statistical analysis showed no significant difference between both systems. The observed percentage of phosphorus removals were considerably lower than the removals obtained in previous experience (Caicedo *et al.*, 2002) due basically the higher influent phosphorus concentration of the wastewater used in this study.

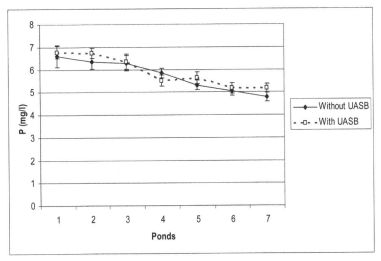

Fig. 7. Effluent Phosphorus in duckweed pond systems fed with de-gritted or with anaerobically pre-treated domestic wastewater (n =9). Standard errors are indicated.

Faecal coliform removal.
Although investigation of pathogen removal was not a main objective of this study, one sample was taken once a month from the influent and effluents. Geometric means are presented in Table 2. In the system without UASB reactor the FC removal was 3.17 log units. In the system with UASB reactor the FC removal was 0.86 log units in the UASB reactor and 2.69 log units in the pond system.

Table 2 Geometric mean values of FC (CFU/100 ml) in the influent and effluent of each system (n = 3).

Description	Values
De-gritted wastewater	4.6×10^6
Effluent system Without UASB reactor	3.1×10^3
Effluent of UASB reactor	6.3×10^5
Effluent of system with UASB reactor	1.3×10^3

Biomass Production
The duckweed biomass production rate in the system receiving de-gritted wastewater was in the range of 36-84 g m^{-2}-d^{-1} (1.75 - 4.1 g m^{-2}-d^{-1}, dry weight)

(Fig. 8). In the system with UASB reactor the production rate was in the range of 54-90 g m^{-2}-d^{-1} (2.5 – 4.2 g m^{-2}-d^{-1}, dry weight). It has been observed repeatedly that the biomass had growth difficulties in the first two ponds of this system without pretreatment and in the first pond of the system with pre-treatment. Protein content was 35.1% and 36.6% for the system with and without pre-treatment respectively. The overall biomass productions were 138 ± 18 g d-1 and 122 ± 17 g d-1 for the system with and without anaerobic pre-treatment respectively, and they were not significantly different at 95% but at 92% level of confidence.

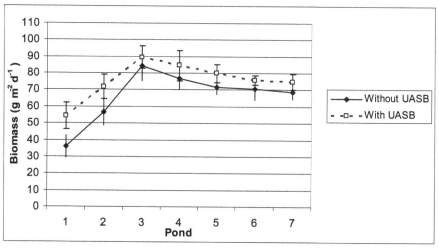

Fig. 8 Biomass Production in duckweed pond systems fed with de-gritted or with anaerobically pre-treated domestic wastewater (n =20). Standard errors are indicated.

Discussion

The Ginebra wastewater can be considered as a typical domestic wastewater of medium concentration (Metcalf and Eddy, 1991). The concentration of ammonia and ammonium nitrogen are not expected to inhibit duckweed growth since ammonium nitrogen concentrations was considerably lower than 50 mg l^{-1} and pH was very close to neutral (Caicedo et al., 2000).

In the system without pretreatment the organic matter is progressively removed along the system. The range of organic loading rate varied between 591 Kg BOD$_5$ ha^{-1}-d^{-1} in the first pond and 50 Kg BOD$_5$ ha^{-1}d^{-1} in the last pond. Clearly the first pond worked as a low loaded anaerobic pond (83 g BOD$_5$ m^{-3} d^{-1}) and the second as a facultative pond with an average load of 245 Kg BOD$_5$ ha^{-1}d^{-1} (Mara et al., 1992; Metcalf and Eddy, 1991). In the system with pre-treatment the organic loading ranged between 258 and 61 Kg BOD$_5$ ha^{-1}d^{-1}.

In Colombia, a wastewater treatment plants are usually designed to fulfill the requirements of Water Authorities according to environmental conditions on each region. In Valle-Colombia, the region where this experiment was run, the

Corporation Autónoma Regional del Valle del Cauca (CVC) requires secondary treatment for the big cities with a minimum removal of 85% in BOD_5 and 90% in total suspended solids, which means in this case, an effluent with 38.1 mg BOD_5 l^{-1} and 16 mg TSS l^{-1}. The BOD_5 requirement was reached in Pond 4 (HRT=12 d) for both systems and TSS requirement was reached in pond 2 (HRT=6 d) in both systems. For small cities, CVC can lower down the requirements depending on technical criteria. If a minimum requirement of 80% is set up, it will correspond to an effluent concentration of 51 mg BOD_5/l. In the system without pre-treatment, this requirement is still reached in pond 4 (HRT=12 d), while in the system with pretreatment, an 80% reduction is reached already in pond 1 (HRT=3.3 d, including retention time in the UASB reactor), This means that the same result can be achieved in the system with UASB in almost a quarter of the area required in the system without pre-treatment. Both systems were able to produce very good effluents in terms of organic matter and suspended solids (BOD_5< 30 mg l^{-1}l, SS< 30 mg l^{-1}, American Standard for stabilization ponds). The results for organic matter and suspended solids are comparable to other studies conducted on domestic wastewater treatment with duckweed stabilization ponds (Zirschky and Reed, 1988; Mandi, 1994; Alaerts et al., 1996; Al-Nozaily, 2001; Zimmo et al., 2002). Effluents of duckweed systems are usually very clear because the duckweed matt provides excellent quiescent conditions for settling and prevents algae growth within the system.

Faecal Coliforms overall removal efficiencies were 95% for the system without pretreatment and 97.9 % for the system with pretreatment. The faecal coliform decay is usually considered a first-order process (Marais, 1974). Using a first order equation for a series of ponds, the faecal coliform die-off coefficient (K_b) was 0.61 d^{-1} for the system without pre-treatment and 0.47 d^{-1} for the system with pre-treatment. Lower Kb values in the system with pretreatment may be explained by the lower TSS concentration in the influent to this system, which results in less removal of FC via attachment and settling. The die-off duckweed ponds coefficient shows a large variation in the literature: 0.3-1.2 d^{-1}, 20 °C, Buuren and Hobma (1991), 0.7-3.2 d^{-1}, 18 -31 °C (Steen et al., 1999), 0.24-0.25 d^{-1} (Vroom and Weller, 1995), 0.16-0.45 d^{-1} in warm season and 0.09-0.14 in cold season (Zimmo et al., 2002). This variation is due to the very complex physical, chemical and biological processes that interact during the removal of FC in a stabilization pond, like natural decay, predation, adsorption to solids and sedimentation.

The UASB reactor removed small percentage of nitrogen (4.1%). It has effect on hydrolysis of organic nitrogen while in the system without UASB the same process occurs at a lower rate, mainly in the anaerobic sediments. Alaerts et al. (1996) reported hydrolysis of organic nitrogen as a limiting step for enhanced biomass production. This was not the case in this experience because even though there was still organic nitrogen to be hydrolyzed in all ponds, sufficient ammonium nitrogen was available for biomass growth, while in their experience the ammonium nitrogen was exhausted in the last part of the treatment system.

The level of significant difference for overall biomass production was not found at the 95% level but the 92%, but biomass production was always higher in the system with pre-treatment. A general observation was that biomass in the first ponds of the system with UASB reactor looked healthier than the biomass in the other system. Both systems have the best production in the middle ponds (3, 4 and 5). The lower production in the first two ponds may be related to high organic loading rates (above 245 Kg BOD_5 $ha^{-1}d^{-1}$) and in the last two ponds to low organic loading rates or deficiency of some micronutrients (Fig. 8). A wide variation of biomass production and protein content is found in literature due to different factors like duckweed specie, nutrient concentrations, organic loading rate, environmental conditions, etc. The biomass production obtained in this study was comparable with the values reported for *Spirodela polyrrhiza* by Sutton and Ornes (1977): 1.9 g m^{-2}-d^{-1} (dry weight), Oron *et al.* (1986): 1.27 – 12 g m^{-2}-d^{-1} (dry weight)), Edwards *et al.* (1992): 2.65 – 5.9 g m^{-2}-d^{-1} (dry weight), Alaerts *et al.* (1996): 1 – 5.5 g m^{-2}-d^{-1} (dry weight). Protein content was in the middle range of the values found in literature (Sutton and Ornes, 1977; Landolt and Kandeler, 1987; Oron *et al.*, 1986; Edwards *et al.*, 1992; Vermaat and Hanif, 1998).

The concentration of phosphorus in the effluent of both systems is higher than the EU standard for effluent of wastewater treatment plants. The removal of phosphorus from aquatic environments is a difficult problem to solve. Due to the role it plays in biological processes its consumption is not as high as for nitrogen. Also the pathways through which it can be removed in its natural cycle are more limited, and basically results in accumulation in the sediment. The reuse of effluents in irrigation is an alternative to obtain further phosphorus removal from the water and to recover it in the form of crop products.

Both systems produced effluent total nitrogen close to the EU standard for nitrogen (TN< 10), with ammonium nitrogen concentrations below this value. Part of the nitrogen was recovered as high quality protein that can be used as a high quality animal feed, while the final effluent still have some nitrogen to be used in irrigation for further recovery in agricultural products. An increasing accumulation of oxidized nitrogen forms (nitrite and nitrate) were observed after pond 4 in both systems, with significant higher values in the system with pretreatment. The reason for this is not clear, as the oxidized nitrogen concentration in the effluent is a result of the couple process nitrification-denitrification. The accumulation trend in both system maybe an indication of an increase of nitrification rates along the systems but it could also be a result of a decreased in denitrification rate. Pre-treatment may enhance oxygenation and nitrification in the ponds and at the same time it may reduce denitrification because organic matter is necessary as substrate for this process. Zimmo (2003) found that nitrification and denitrification rates have a constant tendency along a duckweed ponds system with 4 ponds in series and 28 days of retention time. Further investigation on nitrogen balance on the systems is needed in order to get better understanding of the nitrogen transformations in the systems.

Conclusions

The effect of anaerobic pretreatment on the performance of duckweed based stabilization ponds was studied in two identical systems with seven ponds in series and a retention time of 21 days. No effect of anaerobic pretreatment was observed on pH and temperature in the two systems. Oxygen concentrations were higher in the system with UASB reactor. In terms of organic matter removal, the pretreatment with UASB reactor may contribute to the reduction of area requirement.

Total nitrogen removal was similar in both systems (64% and 68% in lines with and without pretreatment, respectively), with slightly better results in the system without pretreatment. In the system with UASB, higher oxygen concentration and lower organic matter concentration are present which gives better conditions for nitrification while in the system without UASB reactor, higher organic matter concentration may enhance denitrification.

The decision to include a UASB reactor or not depends on the main objective of the wastewater treatment. If the objective is the removal of organic matter and solids, the UASB reactor reduces considerably the area requirement. If the objective is to remove nitrogen, this can be achieved via denitrification or via plant biomass uptake. The system with UASB reactor seems to enhance biomass production and as a consequence nitrogen removal via biomass uptake. Reuse of this biomass in animal feeding or fish aquaculture could generate incentives for wastewater treatment.

It is important to understand how the different nitrogen removal mechanisms contribute to the overall removal and how these can be influenced. This information will be useful to optimize the removal strategy in the direction of more denitrification or more plant biomass production. Further research will be focused on understanding nitrogen transformations and removal mechanisms in duckweed covered sewage stabilization ponds.

Acknowledgements
The authors would like to thank Mr. Luis Felipe Saavedra his help in the collection of the data. The authors thank Acuavalle for facility support and the Netherlands Government and Universidad del Valle for their financial support via the cooperative project ESEE, which is funded by the SAIL program.

References
Alaerts G., Mahbubar Rahman, Kelderman P. (1996). Performance Analysis of a full-scale duckweed-covered sewage lagoon. *Wat. Res.* 30 (4), 843-852.

Al-Nozaily F. A. (2001). Performance and Process Analysis of Duckweed-Covered Sewage Lagoons for high Strength Sewage. Doctoral Dissertation. Delft University of Technology- International Institute of Hydraulic and Environmental Engineering. Delft-Holland.

A. P. H. A. (1995). American standard methods for the examination of water and wastewater. 19th edition. New York.

Buuren J. van and Hobma S. J. (1991) The feacal coliform die-off at post treatment fo anaerobically pretreated wastewater. Internal Report. Department of Environmental Technology, Wageningen Agricultural University, The Netherlands

Caicedo J. R., Steen P. van der, Arce O., Gijzen H. (2000). Effect of total ammonium nitrogen concentration and pH on growth rates of duckweed (*Spirodela polyrrhiza*). *Wat. Res.* 34(15), 3829-3835.

Caicedo J.R., Espinosa C., Gijzen H., Andrade M. (2002). Effect of anaerobic pretreatment on physicochemical and environmental characteristics of Duckweed based ponds. *Wat. Sci.Tec.* 45(1), 83-89.

Daniel W. W. (1990). Applied nonparametric statistics. Georgia State University. Houghton Mifflin Company. Boston.
Corporación Autónoma Regional del Valle del Cauca (CVC). (1976). Acuerdo No. 014 del 23 de Noviembre de 1976.

Edwards P., Hassan M.S., Chao C.H., Pacharaprakiti C. (1992). Cultivation of duckweed in septage-loaded earthen ponds. *Jour. Bioresource Tech.*, 40, 109-117.

Landolt E, Kandeler R. (1987). The family of *Lemnaceae* monographic study, Vol.2. *Veroeffentlichungen des geobotanisches Institutes der ETH*, Stiftung Rubel, 95, Zurich, 1-638.

Mandi L. (1994). Marrakesh wastewater purification experiment using vascular aquatic plants Eichornia *crassipes* and *Lemna gibba*. *Wat. Sci. Tech.* 29(4), 283-287.

Mara D. D., Alabaster G. P., Pearson H. W., Mills S. W. (1992). Waste stabilization ponds. A design manual for Eastern Africa. Lagoon Technology International. Leeds, England.

Marais G.v. R.(1974). Faecal bacterial kinetics in stabilization ponds. Jour. Environ. Eng. Div., ASCE. 100: 119-139.

Metcalf and Eddy. (1991). Waste Engineering. Treatment, disposal and reuse. Tchobanoglous G. and Burton F. L. [eds.]. 2nd Ed. McGraw Hill, Inc. USA.

Oron G., Porath D., Wildschut L.R. (1986). Wastewater treatment and renovation by different duckweed species. *Jour. Environ. Eng. Div., ASCE.* 112(2), 247-263.

Reddy K.R. and DeBusk W.F. (1985) Nutrient removal potential of selected aquatic macrophytes. *Jour. Environ. Qual.*, 14(4), 459-462.

Skillicorn P., Spira W., Journey W. (1993). Duckweed aquaculture, a new aquatic farming system for developing countries. *The World Bank.* 76 p. Washington.

Steen P. van der, Brenner A., Buuren J. van, Oron G. (1999). Post-treatment of UASB reactor effluent in an integrated duckweed and stabilization pond system. *Wat. Res.* 33(3), 615-620.

Sutton D.L. and Ornes W. H. (1977). Growth of Spirodela polyrhiza in static sewage effluent. *Aquat. Bot.* 3, 231-237.

Vermaat J.E., Hanif M. K. (1998). Performance of common duckweed species (*Lemnaceae*) and the waterfern *Azolla filiculoides* on different types of waste water. *Wat. Res.* 32 (9), 2569-2576.

Vroom R. and Weller B. (1995). Treatment of domestic wastewater in a combined UASB-reactor duckweed pond system, Doktoraal verslagen, serie Nr 95-07, Dept. Env.Tech. Wageningen Agricultural University, The Netherlands.

Zimmo O., Al-Sa'ed R. M., Steen P. van der, Gijzen H. (2002). Process Performance Assessment of algae-based and duckweed-based wastewater treatment systems. *Wat. Sci. Tech.* 45(1).

Zimmo O. (2003). Nitrogen transformation and removal mechanism in algal and duckweed waste stabilization ponds. Ph. D. Dissertation. Wageningen University and International Institute of Hydraulic and Environmental Engineering. Holland.

Zirschky J. and Reed S. (1988). The use of duckweed for wastewater treatment. *Jour. Wat. Poll. Cont. Fed.* 60 (7), 1253-12.

Chapter 5

Nitrogen balance of duckweed covered sewage stabilization ponds

Chapter 5

Nitrogen balance of duckweed covered sewage stabilization ponds

Abstract
Nitrogen removal is nowadays one of the most important effluent treatment objectives because of the serious pollution problems it causes to the environment. The conventional processes for wastewater treatment, including chemical and biological removal of nutrients are costly in terms of investment and operation. Duckweed covered sewage stabilization ponds offer a low cost solution to both TSS, COD, pathogen and nutrient removal and includes the possibility of nutrient recovery and re-use. It is important to study how nitrogen is transformed and removed in duckweed ponds. There are not many results available in literature that could be useful to optimize pond design for nitrogen removal and recovery. How much nutrients should be removed may differ depending on the end use of the final effluent. In case of effluent irrigation, residual N could be useful, while in case of discharge levels should be as low as possible.

The experiment was performed in a pilot plant, which consisted of two lines of seven duckweed ponds in series. One line received de-gritted domestic wastewater. The other line received effluent of a real scale Up-flow Anaerobic Sludge Blanket (UASB) Reactor. Both lines were operated at a hydraulic retention time of 21 days, pond depth of 0.70 m, and harvesting interval of 4 days. Nitrogen balances were established for every pond and for every line.

Ammonia volatilization is not an important removal mechanism in duckweed ponds (less than 1%). Removal by sedimentation was also low at 2.1% and 4.7% for the systems with and without anaerobic pre-treatment respectively. Instead, denitrification was found to be the most important removal mechanism (42% and 48 %), followed by duckweed biomass up-take (15.6% and 15.1%. Average nitrogen biomass up-take rates were 199 mg N m^{-2} d^{-1} and 193 mg N m^{-2} d^{-1} for the system with and without pre-treatment, respectively. Nitrification rates were in the range of 112 – 1190 mg N m^{-2} d^{-1} and 58-1123 mg N m^{-2} d^{-1} for the system with and without anaerobic pretreatment respectively. Denitrification rates were in the range of 112 – 937 mg N m^{-2} d^{-1} and 59 – 1039 mg N m^{-2} d^{-1} for the system with and without pre-treatment respectively. The configuration of the system, in particular the down and up flow pattern seems to have an important positive effect on denitrification rates.

Key words
Ammonium volatilization, anaerobic pretreatment, denitrification, domestic wastewater treatment, duckweed ponds, *Lemnaceae*, nitrification, nitrogen mass balance, nitrogen removal.

Introduction
Nitrogen removal is nowadays one of the most important objectives of wastewater pollution control, because of the serious pollution problems it causes to the environment. Ammonium nitrogen is known to be toxic to aquatic life and exerts an

oxygen demand on the water. In the form of nitrate, it may cause infant methaemoglobinemia if present in the water supplies or eutrophication if present in surface waters. To remove nitrogen from wastewater, both conventional and non-conventional systems can be used. The conventional processes for wastewater treatment, including chemical and biological removal of nutrients are costly in terms of investment and operation. In past decades, much research has therefore been done on other technologies for wastewater treatment such as constructed wetlands, land application, aquaculture and reuse, algal ponds (Zirschky and Reed, 1988; Brix and Schierup, 1989; Skillicorn, 1993; Al-Nozaily, 2001) and others. Duckweed stabilization ponds are among these new technologies (Zimmo, 2003a). These are low cost systems, which are also used for nutrient recovery into animal feed.

Nitrogen transformations in wastewater treatment systems depend on the conditions in the aquatic medium such as pH, oxygen, temperature, organic matter content, concentration of different nitrogen compounds and the type of microorganisms present in the system (Metcalf and Eddy, 1995; Bitton, 1999). These parameters are related to the type of treatment system, type of plant configuration, operational conditions and, in the case of natural systems, the type of plants used. Organic nitrogen is hydrolyzed and converted to ammonium nitrogen, which then may be subject to different transformations: heterotrophic assimilation for synthesis and growth, oxidation by autotrophic nitrifying bacteria, assimilation by autotrophic microorganisms and plants, or it may be volatilized depending on the pH conditions of the system (Caicedo et al., 2000). Nitrification is performed in two steps by autotrophic microorganisms. It is generally attributed to *Nitrosomonas europea* and *Nitrobacter agilis*. These bacteria are aerobic, which means that they only grow in the presence of oxygen, but the absence of oxygen is not lethal to them (Painter, 1970). A variety of organic and inorganic agents can inhibit the growth and action of these organisms (Metcalf and Eddy, 1991; Water Environment Federation, 1998). Denitrification can be a biological or a chemical process but the biological one is the most important in natural systems. Several heterotrophic microorganisms perform denitrification under anoxic conditions. It requires an electron donor such as organic matter or other reduced compounds like sulphide or hydrogen ions (Mosier and Klemendtsson, 1994). Plants usually use nitrate as a source of nitrogen but some aquatic plants, like duckweed, prefer ammonium nitrogen and only use nitrate when ammonium is absent from the medium (Porath and Pollock, 1982).

In conventional pond systems, the most common processes reported to be responsible for nitrogen removal are ammonium volatilization (Pano and Middlebrooks, 1982; Silva et al., 1995), denitrification and sedimentation (Zimmo et al., 2003a). The first one depends on the hydrolysis of organic nitrogen and on pH conditions. The second depends on the presence of oxidized nitrogen, DO and on availability of organic matter as electron donor (Pano and Middlebrooks, 1982). The third depends on the content of settling solids and algae in the system.

It is important to study how nitrogen is transformed and removed in duckweed ponds. There are not many results available from literature that could be useful to optimize pond design for nitrogen removal and recovery. The environmental characteristics of the water phase in duckweed stabilization ponds are different from those in conventional stabilization ponds. Therefore, it is expected that nitrogen transformations and removal will also be different. Anaerobic pre-treatment significantly changes the concentration and composition of organic matter and as a consequence the conditions that affect nitrogen transformation. The effect of inclusion of an anaerobic pre-treatment stage on nitrogen transformation processes was evaluated in this study.

Materials and methods.

Experimental set up
The experiment was performed in a pilot plant, which consisted of two lines of seven duckweed ponds in series. One line received de-gritted domestic wastewater. The other line received effluent of a real scale Up-flow Anaerobic Sludge Blanket (UASB) Reactor treating the same domestic sewage (Fig. 1). The UASB Reactor has a volume of 280 cubic meters and a retention time of 7-8 hours. Each duckweed pond consists of a plastic cylinder tank of 0.9 m height and 0.26-m^2 surface area.

The raw domestic wastewater originates from Ginebra, a small municipality in Colombia of about 8.000 inhabitants. The village is located in the Valle del Cauca Province and has a tropical climate. The medium ambient daily temperature during the experiment was 22.2 °C, with a maximum of 32 °C and a minimum of 16 °C.

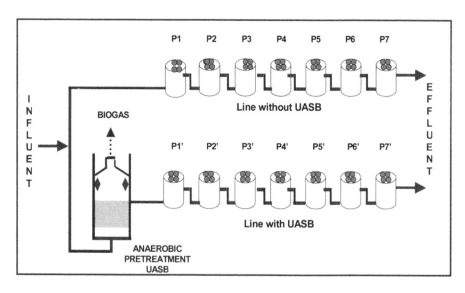

Fig. 1. Schematic diagram of experimental system. P1 to P7 stands for ponds 1 to 7 in the line without pretreatment. P1' to P7' stands for pond 1' to 7' in line with pretreatment.

Pond operational conditions
The duckweed system was operated with a continuous flow of 42 ml min^{-1} and a hydraulic retention time (HRT) of 3 days for each tank, with a total HRT of 21 days.

The pilot plant had been operated during a year before starting the monitoring under this experiment. This experiment lasted for three months. The water level of each tank was kept at 0.70 m height; the inlet was at 20 cm from the bottom and the outlet is at 10 cm under the water surface to avoid loss of duckweed. Ponds were covered with duckweed (*Spirodela polyrrhiza*).

Part of the duckweed cover was harvested every four days. Biomass samples were taken with a strainer of a known area, where it was allowed to drain for about 5 minutes. The fresh weight was determined and the density was calculated. Enough biomass was harvested to leave a density of 700 g m^{-2} (fresh weight) after each harvesting. Previous experiments showed that this density prevents algae growth. The data of biomass production were grouped for every two harvests and production was calculated in g d^{-1} for each pond. The production in each system was the sum of the production in the individual ponds.

The daily average concentrations of different forms of nitrogen of the de-gritted sewage and UASB effluent were determined by averaging the results from twelve 24-hour sampling programs. In each sampling program, the day was divided in periods of four or six hours. A composite sample was collected during each period by taking every half an hour a fixed volume of sample. Grab samples were taken from each pond effluent once a week and all standard nitrogen analysis were performed.

Analytical methods

Chemical Analysis. Total Kjeldahl Nitrogen (TKN), Ammonium Nitrogen (NH$_4^+$-N), Nitrate Nitrogen (NO$_3^-$-N) and Nitrite Nitrogen (NO$_2^-$-N) were measured according to the American Standards Methods (APHA, 1995).

Ammonia Volatilization. For the determination of ammonia volatilization, a method described by Shilton *et al.* (1995) and Zimmo *et al.* (2003b) was used with some modifications. The headspace above the pond (0.26 m^2) was closed with a plastic cover and a constant airflow was evacuated through this space using a vacuum pump. The pump ran continuously (during 24 hours) to ensure the airflow across the pond surface. The pumped air was forced through a column and a conical flask filled with a boric acid solution (2%) where the ammonia was trapped. After 24 hours, the boric acid was titrated with standard 0.02 N H$_2$SO$_4$ to calculate the amount of N-NH$_3$ trapped per day. This value was divided by the surface area of the pond to obtain the amount of volatilized ammonia in mg m^{-2} d^{-1}. Figure 2 shows a diagram of the experimental set-up. This determination was performed for every pond once a month.

Sediment Sampling and Analysis. Part of nitrogen is accumulated in the bottom of the system by sedimentation of solids. This was evaluated by measuring the rate of solids accumulation and its nitrogen content.

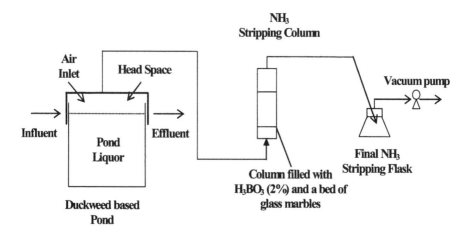

Fig. 2. Diagram of experimental set-up used for measurement of ammonia volatilization.

The rate of solids accumulation was determined by measuring the depth of sludge at the beginning and at the end of the experiment. To determine the depth of sludge a modification of the method proposed by Oostrom (1995) was used which consists of introducing a wooden or plastic stick covered with absorbent cotton lining into the bottom of each pond. Once it reaches the bottom, it should be rotated several times to allow the sediments to stick to the cotton cloth. The stick is taken out carefully and the depth of sludge is measured with a ruler.

Samples of sludge were taken from the ponds with a special device, which opens and closes right on the bottom to avoid sludge dilution. The samples were analyzed to determine the nitrogen content in triplicate.

Duckweed Sampling and Analysis. Every two weeks, the harvested biomass from each pond was mixed and 20 grams of fresh weight (FW) were dried in the oven at 70°C for 24 hours to determine the dry weight. The dry matter was powdered in a grinder and nitrogen content was determined. The nitrogen up-take by the biomass is calculated from the fresh weight biomass production in the system multiplied by its % dry weight and by its percentage of nitrogen content.

Nitrogen Balance
The law of conservation of matter was used as the basis for mass balance in this study. The mass balance equation that was used is the following:

$$N_{in} = N_{out} + N_{UASB} + N_{dw} + N_d + N_v + N_s - N_f \qquad (1)$$

Where,

N_{in} = $(N_{kj} + NO_3^- + NO_2^-)$ in the influent
N_{out} = $(N_{kj} + NO_3^- + NO_2^-)$ in the effluent
N_{kj} = Kjeldahl nitrogen (Organic nitrogen + ammonia nitrogen)
N_{UASB} = N removed in the UASB Reactor
N_{dw} = N removed via duckweed harvesting
N_d = N removed via denitrification
N_v = N removed via ammonia volatilisation
N_s = N-accumulation in sediment
N_f = N-fixation

All fluxes are expressed in g d $^{-1}$. In equation (1), the denitrification flux was assumed as the nitrogen input minus all other nitrogen outputs. The nitrification flux was calculated as: denitrification + effluent oxidized nitrogen.

Data analysis
The data were compared and analyzed statistically; results for different treatments lines were compared using ANOVA test and non-parametric Krusk and Wallis method (Daniel, 1990). The first methods assume independency on the compared data while the second method does not. The results obtained by the two methods were very similar. The results within each treatment line were compared using the pair t-student test.

Results

Nitrogen content in the raw wastewater
The characterization of the de-gritted wastewater and the effluent of the UASB Reactor in terms of the different forms of nitrogen are presented in Table 1.

Table 1. Nitrogen content of the de-gritted wastewater and UASB effluent.

Parameter	Units	De-gritted wastewater *	UASB effluent*
TKN	mg l^{-1}	38.2 ± 4.7	36.8 ± 5.5
N-Ammonium	mg l^{-1}	26.4 ± 4.4	30.5 ± 5.3
N-Organic	mg l^{-1}	11.8 ± 4.4	6.3 ± 2.2
N-Nitrites	mg l^{-1}	0.04 ± 0.02	0.02 ± 0.01
N-Nitrates	mg l^{-1}	0.04 ± 0.02	0.02 ± 0.01

*Averages ± S.D. (n=12 daily average; each n = average of 4 or 6 integrated samples/day)

Nitrogen removal pattern in the systems
The composition of the different forms of nitrogen (Ammonium- N, Organic- N, Nitrate-N and Nitrite-N) in the influent and effluents of ponds for both systems, is shown is Figure 3.

Nitrogen Balance

To apply equation (1) it was necessary to establish the different nitrogen fluxes in the two systems.

The results of ammonia volatilization for each pond were added all together in each treatment line. The average results were 10.3 ± 0.7 mg N d^{-1} and 9.7 ± 0.6 mg d^{-1} for the system without and with pre-treatment, respectively (n = 21; 7 ponds, 3 measurements/pond). This is less than 1% of influent nitrogen.

The accumulation of nitrogen in the sediments was 108 ± 1 mg N d^{-1} and 50 ± 1 mg N d^{-1} (n = 21; 3 measurements/pond, 7 ponds) in the systems without and with pre-treatment respectively.

The results for biomass production, dry mass, nitrogen content and nitrogen biomass up-take are presented in Table 2.

Influent and effluent nitrogen loads were obtained by multiplying the flow with the total nitrogen concentration [$NH_4^+ + $ N Org $ + NO_2^- + NO_3^-$]. The effluent flow was compared with the influent flow along the day during several days. Only during the sunny hours, the effluent flow was slightly lower than the influent flow, but this represented less than 1% of average flow. No precipitation was taken into consideration as the systems were protected from the rain.

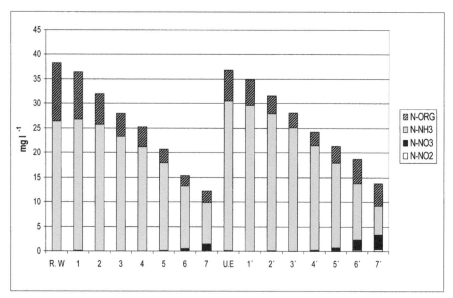

Fig. 3. Concentration of different nitrogen forms in the influent and pond effluents of the system without UASB Reactor (left) and of the system with UASB Reactor (right). RW stands for raw wastewater, UE for UASB effluent.

Table 2. Nitrogen biomass up-take for systems fed with de-gritted wastewater and UASB effluent

Parameter	Units	System without UASB	System with UASB
Biomass production (fresh weight)	g d^{-1}	122 ± 17 (n=18)	138 ± 18 (n=18)
Dry mass	%	4.90 ± 0.1 (n=6)	4.66 ± 0.1 (n=6)
Nitrogen content of dry mass	%	5.86± 0.1 (n=10)	5.62 ± 0.1 (n=10)
Nitrogen biomass up-take	g d^{-1}	0.35	0.36

Nitrogen fixation is assumed absent because there is presence of ammonium nitrogen in the systems along the seven ponds. Brock *et al.* (1991) reported that in the presence of ammonia, nitrogenase synthesis is suppressed by a phenomenon called the "ammonia switch-off" effect. On the other hand, Duong and Tiedje (1985) reported N-input in naturally occurring duckweed-cyanobacterial associations of 1-2 mg N m^{-2} d^{-1}. Considering this value as possible input of nitrogen and comparing it with the other nitrogen fluxes in the system, it is concluded that the potential contribution of N-fixation to the N-mass balance is negligible.

In equation, (1) all nitrogen fluxes have been measured, but the denitrification flux is calculated as the difference between the nitrogen inputs minus all the other effluent nitrogen fluxes in the systems. The results of the mass balance per pond are shown in Figure 4. The overall nitrogen balance is presented in Figure 5.

Discussion

The total nitrogen removal efficiencies were 64 % and 68 % in the system with and without anaerobic pretreatment, respectively. Some remaining nitrogen in the effluent can be useful if it is going to be re-used in irrigation. The adequate effluent nitrogen concentration will depend on the type of crop to be irrigated. The effluent total nitrogen concentrations in both lines (Fig. 3) were fairly close to the EU standard for effluent of wastewater treatment plants (10 mg N l^{-1}). If the effluent should be discharge to surface waters it is necessary to improve the removal efficiency.

The mass balance showed that ammonia volatilization is not an important removal mechanism in duckweed based stabilization ponds (Fig. 5). It represented only 0.4-0.5% of influent nitrogen in both systems. Not many data are found in literature for direct measurement of this parameter. These percentages are within the same range as those reported by Zimmo *et al.*, (2003b) for duckweed ponds. Furthermore, the same authors found that, despite of higher pH values, even in conventional stabilization ponds this mechanism does not play an important role in nitrogen removal. Ferrara and Avci (1982) arrived at similar conclusions when modeling nitrogen dynamics in waste stabilization ponds.

Effect of Operational Variables on Nitrogen Transformations in Duckweed Stabilization Ponds

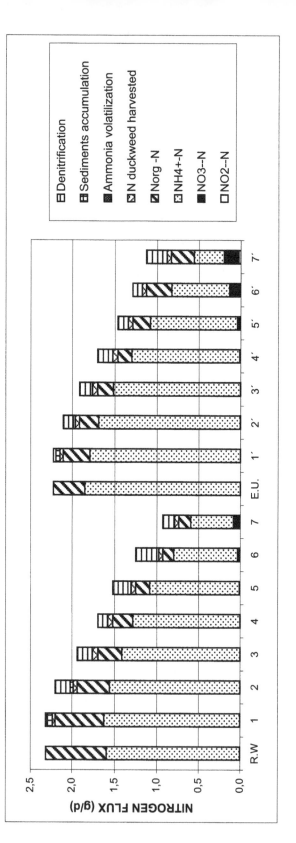

Fig. 4. Nitrogen Balances for each pond in the duckweed systems fed with de-gritted wastewater (left) and with UASB effluent (right). RW stands for raw wastewater, EU for UASB effluent

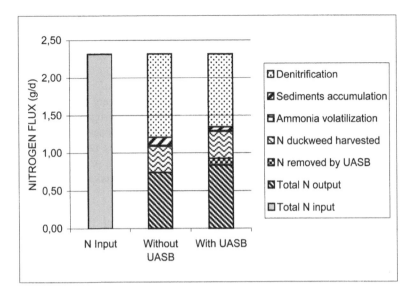

Fig. 5. Overall nitrogen mass balance for the system fed with de-gritted wastewater and the system fed with UASB effluent (g N d^{-1}).

From the nitrogen balance, it can be seen that sedimentation was not a predominant mechanism for nitrogen removal (2.1 and 4.7% for the systems with and without UASB reactor, respectively). Influent total suspended solids concentrations were 62 ± 8 mg TSS l^{-1} and 147 ± 23 mg TSS l^{-1} for the system with and without UASB reactor, respectively. On the contrary, Ferrara and Avci (1982), using a modeling program, concluded that sedimentation was the predominant mechanism for N-removal in algae ponds and Zimmo et al. (2003a) reported sedimentation as the second most important nitrogen removal mechanism (10–22%), with influent suspended solids concentration in the range of 149-189 mg TSS l^{-1}. The accumulation of nitrogen in the bottom of the systems will depend on different factors like suspended solid concentration, anaerobic biodegradability of the settled solids and hydraulic reactor characteristics. In the system without pre-treatment 95 % of the nitrogen accumulated in the solids was settled in the first two ponds and in the system with pre-treatment 97 % of the nitrogen accumulated in the solids was settled in the first pond. The accumulation of solids in the rest of the ponds was very low, which means that the contribution of dead duckweed biomass to the nitrogen removal on the sediments was not important. The reason for this may be that short harvesting periods (4 days) produce a young viable duckweed mat with a small fraction of decaying plants. This is in agreement with Eighmy and Bishop (1989) who found that the importance of sedimentation mechanism decreased when the plant matt was kept high productive by frequent harvesting. They reported 6% of nitrogen removal by sedimentation. In duckweed ponds, the generation and subsequent settling of algal biomass is low.

The overall nitrogen balance shows that the nitrogen removal via duckweed uptake was around 15%. This was the second most important nitrogen removal mechanism for both systems (Fig. 5). The average up-take rates into the biomass were 199 and 193 mg N m^{-2} d^{-1} in the system with and without UASB reactor, respectively. These rates are in the same range as those reported for *Spirodela polyrrhiza* by Alaerts *et al.* (1996): 260 mg N m^{-2} d^{-1}, Reddy and Debusk (1985): 150 mg N m^{-2} d^{-1} and Kvet *et al.* (1979): 200 mg N m^{-2} d^{-1}. In duckweed systems, nitrogen biomass uptake occurs only in the upper layers of the ponds (Fig. 6). In deeper ponds, a greater proportion of the nitrogen present in the water column has no contact with the plants. This contact may be improved by hydraulic mixing or by using shallow ponds. If the percentage of nitrogen removal via duckweed uptake is to be improved, the availability of area is an important factor.

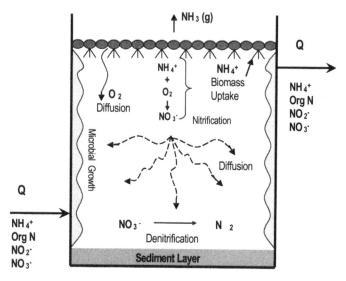

Fig. 6. Diagram of nitrogen mobility in duckweed ponds

Denitrification was the mayor nitrogen removal mechanism for both systems, it accounted for 42 % and 48 % in the systems with and without UASB reactor (Fig. 5). Denitrification rates were in the range of 112 – 936 mg N m^{-2} d^{-1} for the system with pre-treatment and 59 – 1039 mg N m^{-2} d^{-1} for the system without pre-treatment (Fig. 7 and 8). Not many results are found in literature for denitrification rates in duckweed systems. The results reported by Davidsson and Stahl (2000) for a constructed wetland were within the same range found in our experiments (124 mg N m^{-2} d^{-1} for a sandy loam soil and 682 mg N m^{-2} d^{-1} for a peaty soil). Zimmo *et al.* (2003c) reported considerably lower range for denitrification rates of 265-409 mg N m^{-2} d^{-1}, in a pilot scale duckweed system operated at similar ambient temperature and similar domestic wastewater concentration. Vermaat and Hanif (1998) reported also a relative lower denitrification rate (260 mg N m^{-2} d^{-1}) in a duckweed laboratory scale experiment. The compartmentalized configuration of the system, plus the sequencing up-flow pattern may explain the higher denitrification rates obtained in this experiment. The oxidized nitrogen being produced in the upper layer of ponds flowed to the lower anaerobic layer of the next ponds, where better conditions for denitrification were present. Nielsen *et al.* (1990) found a denitrification rate of 386 mg N m^{-2} d^{-1} in biofilms from nutrient rich streams. This was increased to 1075 mg N m^{-2} d^{-1} with the increase of nitrate concentration to 8 mg l^{-1} and to 3024 mg N m^{-2} d^{-1} with addition of organic matter concentration. Dalsgaard and Revsbech (1992) studied the denitrification process in biofilms from trickling filters. The basic denitrificaton rate was 725 mg N m^{-2} d^{-1} at 17.5 mg NO$_3$-N l^{-1}, which was incremented to 3750 mg N m^{-2} d^{-1} with nitrate increase to 175 mg NO$_3$-N l^{-1}. Further addition of organic matter produced even higher denitrification rates (4.838 mg N m^{-2} d^{-1}). In this experiment, the systems have considerable surface area available (Fig. 6) for growth of denitrifiers in the lower layer of the biofilms of pond walls, bottom sediment and duckweed cover to support the observed dentrification rates.

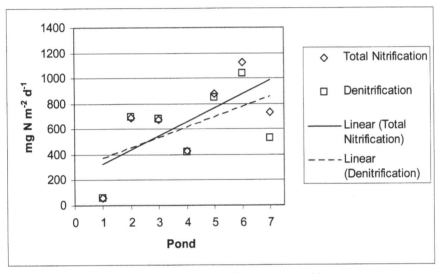

Fig. 7. Nitrification rates and denitrification rates for the system without pre-treatment

Nitrification rates in this study were in the range of 112 -1190 mg N m^{-2} d^{-1} and 58 – 1123 mg N m^{-2} d^{-1} for the system with and without pre-treatment respectively (Fig 7 and 8). These ranges are lower than those reported by Leu *et al.* (1998), who were treating artificial domestic wastewater in an open channel (720 – 2520 mg N m^{-2} d^{-1} for ammonia oxidizers and 360 – 1728 mg N m^{-2} d^{-1} for nitrite oxidizers), perhaps due to the different hydraulic and re-aeration conditions in the channel. The rates found in this study are considerably higher than rates reported by Caffrey *et al.* (1993), for marine sediments. They reported rates of 14 – 39 mg N m^{-2} d^{-1} and 14 – 81 mg N m^{-2} d^{-1} after addition of an intermediate amount of organic matter (15 ^{0}C). These lower rates can be expected in environments with low N contamination. Zimmo *et al.* (2003c) reported lower nitrification rates 286 – 434 mg N m^{-2} d^{-1} for a duckweed pond system, under similar wastewater strength and temperature conditions. They did not observe an increase in nitrification rates along the treatment line, whereas nitrification rates in this experiment did increase with increasing HRT.

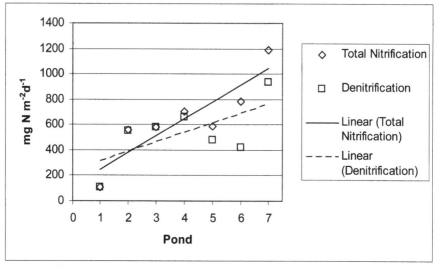

Fig. 8. Nitrification rates and denitrification rates for the system with pre-treatment

The main factors that affect nitrification are temperature, pH, presence of inhibitors, number of nitrifiers and oxygen concentration. The temperature along the system was fairly constant and pH was within the permissible range for nitrifier growth (6.9 - 7.5). Inhibitors were not expected to be present, since the municipality produces only domestic sewage. The number of nitrifiers was not measured but it was expected to be similar as biomass density was kept within the same range in the ponds. Eighmy and Bishop (1989) found equal distribution of nitrifiers in the upper and lower parts of an aquatic macrophyte-based treatment system. Therefore, the factor that may explain the increase in nitrification rates is the increase in dissolved oxygen concentration.

The range of net nitrification rates (nitrification rate – denitrification rate) varied between –5 and 202 mg N m^{-2} d^{-1} along the system without pre-treatment and between –3 and 364 mg N m^{-2} d^{-1} along the system with pre-treatment. The negative values in the first ponds may be an indication of a good potential for denitrification, where denitrification probably was limited by a lower nitrification rate. Eighmy and Bishop (1989) reported comparable net nitrification rates (50 – 200 mg N m^{-2} d^{-1}). Cooper (1983) observed net nitrification rates in the range of 33 – 75 mg N m^{-2} d^{-1} by sediments of a stream receiving geothermal inputs of ammonium at concentrations similar to domestic wastewater. Zimmo et al. (2003c) reported quite lower range of net nitrification rates between 17 and 40 mg N m^{-2} d^{-1} in a duckweed system.

Nitrate accumulation was observed in the last three ponds of the systems (Fig. 3). Even though nitrification and denitrification rates were both increasing, the net nitrification rate was higher in those ponds (Fig. 7 and 8), which indicates that denitrification was not increasing at the same rate as nitrification. Oostrom (1995) demonstrated that 1 mg of NO_3-N requires 2.86 mg COD, to be denitrified. If bacterial synthesis is included, 1 mg of NO_3-N requires 3.7 mg COD, to be denitrified. This organic matter should be biodegradable so it is assumed as 3.7 mg of BOD_5. The amount of BOD_5 needed to denitrify the nitrate still present for the most critical pond (No. 7 in the system with pretreatment) was 0.75 g BOD_5 d^{-1} which was considerably lower than the remaining BOD_5 in the pond (1.75 g d^{-1}). It can therefore be concluded that denitrification was not limited by shortage of organic matter.

Nitrate produced in the upper layers, where oxygen was available, has to be transported to the lower layers to be denitrified. For the pond with the highest nitrate concentration, it was estimated the amount of nitrates diffused to a middle depth of the pond using the expression reported by CRC Handbook (1977). It was concluded that the amount of nitrates transferred from the surface to the lower layer was negligible compared with the available nitrates. So, the diffusion factor can be considered negligible. Apparently, a degree of mixing was present. The more so since ponds 1 to 4 did not show nitrate accumulation.

From above discussion, it can be concluded that the configuration of the system stimulated the accumulation of nitrates in the upper part of the ponds. Part of the nitrates will be transported to the bottom of the next pond or to the effluent of the system. This configuration may have been enhancing denitrification rates in the systems. This is showing the importance of the location of the inlets and outlets of the system, especially when this is compartmentalized. In full scale systems this flow pattern can also be stimulated via the introduction of baffles. This shows that pond design can be largely modified depending on the treatment objectives (stimulate or reduce denitrification). For practical applications, it has to be decided if the interest of the wastewater treatment is to remove nitrogen (stimulate denitrification) or to remove and to recover it in the form of biomass (stimulate duckweed incorporation) and/or application of effluent irrigation (allow residual N in effluent). In the first case, a strategy to improve nitrification-dentrification will be the most useful. In spite of the increasing nitrification rates obtained in the last ponds of the systems, a considerable amount of ammonium nitrogen was still present in the effluents. The introduction of aerobic zones in the first compartments of the system, *i.e.* via intermittent algal ponds, could stimulate nitrification at earlier stages. For full scale application, a baffled system with up and down flow compartments will be recommendable to enhance denitrification.

In the second case, a strategy to improve nitrogen conversion into biomass should be implemented. In this case, the contact of the biomass with the water phase should be enhanced, by altering surface to volume ratios of the system. It would be very interesting to study how the water depth affects nitrogen removal in duckweed ponds. In case of effluent irrigation, crop water and nutrient requirements will dictate the desired residual N content in the effluent. In this case removal mechanisms should be tuned to give the desired level in the effluent.

Conclusions

- Ammonia volatilization is not an important removal mechanism in duckweed ponds. Removal by sedimentation was also low at 2.1% and 4.7% for the systems with and without pre-treatment respectively.

- Instead, denitrification was found to be the most important removal mechanism (42% and 48 % for the systems with and without pre-treatment, respectively), followed by biomass up-take (15.6% and 15.1%).

- Average nitrogen biomass up-take rates were 199 mg N m^{-2} d^{-1} and 193 mg N m^{-2} d^{-1} for the system with and without pre-treatment respectively. Nitrification rates were in the range of 112 – 1190 mg N m^{-2} d^{-1} and 58-1123 mg N m^{-2} d^{-1} for the two systems, while denitrification rates were in the range of 112 – 937 mg N m^{-2} d^{-1} and 59 – 1039 mg N m^{-2} d^{-1} · The configuration of the system, in particular the down and up flow pattern seems to have an important positive effect on denitrification rates.

- Anaerobic pre-treatment had an effect on net nitrification rates of the last three ponds of the system with UASB reactor. Significantly different nitrate concentrations were found in these ponds compared with the system without anaerobic pretreatment.

Acknowledgements
The authors would like to thanks Mr. L.F. Saavedra for his help in the collection of the data, Ms I. Yoshioka for her collaboration in the statistical analysis and the Netherlands's Government and Universidad del Valle for their support in the development of this research through the SAIL ESEE project.

References
Alaerts G., Mahbubar Rahman, Kelderman P. (1996). Performance Analysis of a full-scale duckweed-covered sewage lagoon. *Wat. Res.* 30 (4), 843-852.

Al-Nozaily F. A. (2001). Performance and Process Analysis of Duckweed-Covered Sewage Lagoons for high Strength Sewage. Doctoral Dissertation. Delft University of Technology- International Institute of Hydraulic and Environmental Engineering. Delft-Holland.

A. P. H. A. (1995). American standard methods for the examination of water and wastewater. 20th edition. New York.

Bitton G. (1999). Wastewater Microbiology. 2nd Edition. NY.

Brix H. and Schierup H. H. (1989). The use of aquatic macrophytes in water pollution control. *Ambio.* 18, 100-107.

Brock T. D., Madigan M., Martinko J., Parker J. (1991). Biology of Microorganisms. 17st Edition. Prentice Hall Inc., London.

Caicedo J. R., Steen P. van der, Arce O., Gijzen H. (2000). Effect of total ammonium nitrogen concentration and pH on growth rates of duckweed (*Spirodela polyrrhiza*). *Wat. Res. 34(15), 3829-3835.*

Caffrey J. M., Sloth N. P., Kaspar H. F., Blackburn T. H. (1993). Effect of organic loading on nitrification and denitrification in a marine sediment microcosm. *Microbiol. Ecol.*, 12, 159-167.

Cooper A. B. (1983). Population ecology of nitrifiers en a stream receiving geothermal inputs of ammonium. *Appl. Env. Microbiol.* 45, 1170-1177.

CRC Handbook of Chemistry and Physics. 38th Edition. 1977-1978.

Dalsgaard T. and Revsbech N. P. (1992). Regulating factors of denitrification in trickling filter biofilms as measured with the oxygen/nitrous oxide microsensor. *Microbiol. Ecol.* 101, 151-164.

Daniel W. W. (1990). Applied nonparametric statistics. Georgia State University. Houghton Mifflin Company. Boston.

Davidsson T. E. and Stahl M. (2000). Influence of organic carbon on nitrogen transformations in five wetlands soils. *Soil Science Society of America Journal.* 64 (3). 1129-1136.

Duong T. P. and Tiedie J. M. (1985). Nitrogen fixation by naturally occurring duckweed-cyanobacterial associations. Can. Jour. Microbiol, 31, 327-330.

Eighmy T. T. and Bishop P. L. (1989). Distribution and role of bacterial nitrifying populations in nitrogen removal in aquatic treatment system. *Wat.. Res.* 23 (8), 947-955.

Ferrara R. A. and Avci C. B. (1982). Nitrogen dynamics in waste stabilisation ponds. *Jour. Wat. Poll. Cont. Fed.* 54(4), 361-369.

Kvet J., Rejmankova E., Rejmanek M. (1979). Higher aquatic plants and biological wastewater treatment. The outline of possibilities. Aktiv Jihoceskych vodoh Conf. Pp 9.

Leu H. G., Lee C. D., Ouyang C. F., Tseng H. (1998). Effects of organic matter on the conversion rates o nitrogenous compounds in a channel reactor under various flow conditions. Wat. Res. 32(3), 891-899.

Metcalf and Eddy. (1991). Waste Engineering. Treatment, disposal and reuse. Tchobanoglous G. and Burton F. L. [Eds.]. 2nd Ed. McGraw Hill, Inc. USA.

Mosier A. R. and Klemendtsson L. (1994). Measuring denitrification in the field. Methods soil analysis. Part 2.Microbiologycal and biochemical properties. SSSA Book Series, No. 5.

Nielsen L.P., Christensen P. B., Revsbech N. P. (1990). Denitrification and oxygen respiration in biofilms studied with a microsensor for nitrous oxide and oxygen. Microbial Ecology, 19, 63-72.

Oostrom A. J. van (1995). Nitrogen removal in constructed wetlands treating nitrified meat processing effluent. Wat. Sci. Tech., 32 (3), 137-147

Pano A. and Middkebrooks E. J. (1982). Ammonia nitrogen removal in facultative wastewater stabilisation ponds. Jour. Wat. Poll. Cont. Fed. 54(4), 344-351.
Painter H. A. (1970). A review of literature on inorganic nitrogen metabolism in microorganisms. Wat. Res. 4, 393-450

Porath D. and Pollock J. (l982). Ammonia stripping by duckweed and its feasibility in circulating aquaculture. Aquat. Bot. 13, 125-131.

Reddy K.R. and DeBusk W.F. (1985) Nutrient removal potential of selected aquatic macrophytes. J. Environ. Qual. 14(4), 459-462.

Silva S. A., de Olivieira R., Soares J., Mara D. D., Pearson H. W. (1995). Nitrogen removal in pond systems with different configurations and geometries. Wat. Sci. Tech. 31 (120), 321-330.

Shilton A., Mara D. D., Pearson H. W. (1995). Ammonia volatilisation from a piggery pond. Wat. Sci. Tech. 33 (7), 183-189.

Skillicorn P., Spira W., Journey W. (l993). Duckweed aquaculture, a new aquatic farming system for developing countries. The World Bank. 76 p. Washington.

Vermaat J. E. and Hanif M. K. (1998). Performance of common duckweed species (Lemnaceae) and the waterfern Azolla fliliculoides on different types of wastewater. Wat. Res. 33, 2569-2576.

Water Environment Federation (1998). Biological and chemical systems for nutrient removal. Alexandria, USA.

Zimmo O., Steen N. P van der, Gijzen H. J. (2003a). Nitrogen mass balance over pilot scale algae and duckweed-based wastewater stabilization ponds. In: Nitrogen transformation and removal mechanism in algal and duckweed waste stabilizacion ponds. Ph. D. Dissertation. Wageningen University and International Institute of Hydraulic and Environmental Engineering. Holland.

Zimmo O., Van der Steen N. P, Gijzen H. J. (2003b). Comparison of ammonia volatilisation rates in algae and duckweed-based waste stabilization ponds treating domestic wastewater. *Wat. Res.* 37, 45587-4594

Zimmo O., Van der Steen N. P, Gijzen H. J. (2004c). Quantification of nitrification and denitrification rates in pilot-scale algae and duckweed-based waste stabilization ponds. Env. Tech. 25, 273-282.

Zirschky J., Reed S. (1988). The use of duckweed for wastewater treatment. *Jour. Wat. Poll. Cont. Fed.* 60 (7), 1253-1258.

Chapter 6

Effect of introducing aerobic zones into a series of duckweed stabilization ponds on nitrification and denitrification

Chapter 6

Effect of introducing aerobic zones into a series of duckweed stabilization ponds on nitrification and denitrification

Abstract

Duckweed pond technology presents a low cost alternative for many wastewater treatment applications. Although its potential for removing carbonaceous and suspended material from wastewater has been demonstrated, duckweed systems could be further optimized for nitrogen removal. Nitrogen removal was studied in a pilot plant consisting of seven small duckweed ponds in series. The feed of the duckweed pond system consisted of the effluent of a real scale UASB reactor, which treated domestic wastewater from Ginebra, a small town in Colombia. This experiment was run in two consecutive phases. During the first phase, the seven ponds of the pilot plant were fully covered with duckweed (*Spirodela polyrrhiza).* Before the start of the second phase, the duckweed cover was removed from ponds 1 and 3. The system was operated with a continuous flow to produce a hydraulic retention time (HRT) of 3 days per pond and a total HRT of 21 days. The duration of each phase was three months and the system was monitored for pH, temperature, oxygen and all fluxes of nitrogen.

Effluent total nitrogen was significantly different in the two phases, with 13.8± 2.9 mg TN l^{-1} (63 % removal) and 3.7±1.5 mg TN l^{-1} (90%) for first and second phase, respectively. Denitrification was the most important removal mechanism during phases, and amounted to 43.5 % and 76.2 % of influent nitrogen, in first and second phase, respectively. Ammonia volatilization and sedimentation were insignificant processes for nitrogen removal in both phases.

Nitrification played an important role in nitrogen transformations in the duckweed systems and it was favored by the introduction of aerobic zones. Denitrification also played a key role in nitrogen transformations and removal. Despite the presence of oxygen in the water column, denitrification occurred, probably due to the anaerobic microenvironment of system biofilms. This research showed that higher nitrogen removal might be obtained in duckweed pond systems through the introduction of aerobic zones in early stages of the system. Where effluents can not be reused for crop irrigation, strict nitrogen effluent criteria can be met using hybrid duckweed-algal ponds at considerably shorter hydraulic retention time compared to fully covered systems.

Key words.
Aerobic zones, ammonium volatilization, anaerobic pretreatment, denitrification, domestic wastewater treatment, duckweed, *Lemnaceae*, nitrification, nitrogen removal.

Introduction
In nature, the nitrogen cycle is one of the most complex elemental cycles due to the various oxidation-reduction states nitrogen may have. The level of oxygen in the

environment is one of the most important determining parameters influencing nitrogen transformations.

Duckweed pond technology is being proposed as a low cost alternative for many wastewater treatment applications (Skillicorn *et al.*, 1993;.Alaerts *et al*, 1996; Gijzen, 2001). Although its potential for removing carbonaceous and suspended material from wastewater has been demonstrated (Oron *et al.*, 1986, Oron *et al.*, 1987; Alaerts *et al*, 1996; Al-Nozaily, 2001; Zimmo *et al.*, 2002), duckweed systems usually do not remove all nitrogen within applied HRT of 15 to 20 days. In case effluents are not reused for crop irrigation, duckweed ponds need to be further optimized for nitrogen removal.

Oxygen levels in duckweed ponds are rather low compared to algal systems, due to lack of photosynthetic activity in the water column. Even thought the low oxygen levels, nitrification represented an important transformation route for nitrogen in duckweed ponds (Caicedo *et al.*, 2004). Nevertheless, the presence of ammonium nitrogen concentrations in the system's effluent indicated that the lack of oxygen was a bottleneck for the oxidation of ammonium. Zimmo *et al.* (2003a) also showed that the nitrification process in duckweed ponds is limited due to low dissolved oxygen concentrations. This may suggest that nitrification could be further stimulated by introducing aerobic zones in the system and as a consequence overall nitrogen removal through the couple process nitrification-denitrification.

A similar observation was made by White (1995) who found that passive aeration in the inlet zone of a constructed wetland improved considerably the nitrogen removal efficiency. Grace (1998), working with constructed wetlands, found that the inclusion of open areas at the entrance of the wetland had a positive effect on the nitrogen removal in the system. Also Garcia (1999) reported enhanced oxygen transfer to the water, and consequent higher nitrogen removal, by including areas devoid of vegetation in a constructed wetland treating municipal wastewater.

The hypothesis tested in this research was that low oxygen levels in duckweed systems limit nitrification rates. The inclusion of aerobic zones in a duckweed system might therefore enhance nitrification, and could consequently enhance denitrification processes in the system.

Materials and methods.

Experimental set-up
The pilot plant consisted of seven duckweed ponds in series. The influent was the effluent of a real scale Upflow Anaerobic Sludge Blanket reactor (UASB), which treated municipal wastewater from Ginebra, a small town in Colombia. Average ambient daily temperature during the experiment was 22.2 °C and 22.4 °C, for phase I and II, respectively. The UASB Reactor had a volume of 280 cubic meters and a retention time of 7-8 hours. Each duckweed pond was made out of a plastic cylinder tank of 0.90 m height and 0.26 m^2 area.

Experimental procedures
This experiment was run in two consecutive phases. During the first phase, the seven ponds of the pilot plant were fully covered with a dense layer of duckweed (*Spirodela polyrrhiza*). During the second phase, the duckweed cover from ponds 1 and 3 was removed in order to allow free oxygenation from the atmosphere and from algae photosynthetic activity in the water column (Fig. 1). The experiment lasted for three months per phase.

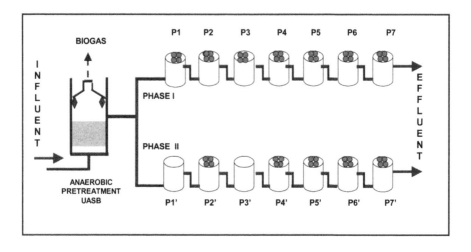

Fig. 1. Diagram of the experimental set-up for the two phases. P1 to P7 series of ponds in phase I, P1' to P7' series of ponds in phase II.

The ponds were operated with a continuous flow of 60.5 l d^{-1} and a water depth of 0.70 m to obtain a hydraulic retention time (HRT) of 3 days per pond and a total HRT of 21 days.

The daily average concentrations of different forms of nitrogen in the influent to the system (effluent of UASB) were determined by averaging the values of twelve 24-hour sampling programs. For each sampling program, the day was divided into periods of four or six hours. Each composite sample was collected during each period by taking every half an hour a fixed volume of sample and adding all together before analysis. In the pond effluents, grab samples were taken between 9 and 11 in the morning once a week and analyzed for nitrogen compounds.

Longitudinal and vertical (depths: 7, 35 and 63 cm.) profiles of pH, temperature and oxygen were made in the system in the morning (8:30 to 9:30) and afternoon (16:30 to 17:30) during the two phases.

The duckweed cover was harvested once every four days. The biomass samples were taken with a strainer of a known area, and were allowed to drain for about 5 minutes. Subsequently, the fresh weight was determined and the density was calculated. Enough biomass was harvested to leave a density of 700 g m^{-2} (fresh weight). Every two weeks, the harvested biomass from the ponds in each system was mixed and 20 grams of fresh weight (FW) were dried at 70°C for 24 hours to determine the dry weight. The dry biomass was grinded and analyzed for nitrogen content.

Physico - Chemical Analysis. Total Kjeldahl nitrogen (TKN), Ammonium Nitrogen (NH_4^+-N), Nitrate Nitrogen (NO_3^--N) and Nitrite Nitrogen (NO_2^--N), Chemical Oxygen Demand (COD), Biochemical Oxygen Demand (BOD_5), Total Suspended Solids (TSS), were measured according to the American Standards Methods (APHA, 1995). Dissolved oxygen, pH, temperature and conductivity were determined with specific electrodes.

Ammonia Volatilization.. For the determination of ammonia volatilization a method described by Shilton *et al.* (1996) and Zimmo *et al.*, (2003b) was used with some modifications. The headspace over the duckweed mat was isolated from the atmosphere. Airflow was created over the water surface of the pond with a pump and the outflow was collected in a boric acid solution during 24 hours with the objective of trapping the ammonia gas evolving from the water surface. The boric acid solution was then titrated to determine the amount of nitrogen in the form of ammonia escaping from the pond. A detailed description of the method is found in Caicedo *et al.* (2003). The measurements were performed once a month for every pond.

Sediment Sampling and Analysis. The nitrogen accumulated in the sediments was evaluated by measuring the rate of accumulation of solids and its nitrogen content. The rate of accumulation of solids was determined by measuring the depth of sludge at the beginning and at the end of the experiment. The depth of sludge was measured by gently introducing a wooden or plastic stick, covered with absorbent cotton lining, to the bottom of each pond. Once it reaches the bottom, it was rotated several times to allow the sediments to stick to the cotton cloth. The stick was taken out carefully and the depth of sludge was measured with a ruler (Oostrom, 1995). Samples of sludge were taken from the ponds with a special sludge sampler, which opens and closes at the bottom of the pond to avoid sludge dilution. The samples were analyzed in triplicate to determine nitrogen content.

Nitrogen Balance.
Nitrogen balances were established for each experimental phase using the following expression, based on the law of mass conservation.

$$N_{in} = N_{out} + N_{dw} + N_d + N_v + N_s - N_f \qquad (1)$$

Where,

N_{in}	$= (N_{kj} + NO_3^- + NO_2^-)$ in the influent
N_{out}	$= (N_{kj} + NO_3^- + NO_2^-)$ in the effluent
N_{kj}	$=$ Organic nitrogen + ammonia nitrogen (Kjeldahl nitrogen)

N_{dw} = N removed via duckweed harvesting
N_d = N removed via denitrification
N_v = N removed via ammonia volatilisation
N_s = N-accumulation in sediment
N_f = N-fixation

All fluxes are expressed in g d $^{-1}$. In equation (1), the denitrification flux was assumed as the nitrogen input minus all other nitrogen outputs. The nitrification flux was calculated as: denitrification + effluent oxidized nitrogen.

Data analysis
The results of different phases were compared using the ANOVA test and the nonparametric Kruskal-Walis test (Daniel, 1990). The latter is a non-parametric method that does not assume independency of the results. The results of the comparison of these two methods were very similar. The results for ponds within each treatment were compared using the paired t-test. Level of confidence was 95%.

Results

Temperature, pH and oxygen profiles. Phase I.
Morning water temperature was between 22 and 23 °C along the system, with no significant differences between ponds at the same depth. In the depth profile small, but significant differences were found between the surface and middle layer (<1 °C). In the afternoon significant differences were found between surface, middle and bottom layer. Temperatures ranged from 27 °C just below the surface to 24 °C close to the bottom.

The pH ranged between 6.6 and 7.4 along the system, with the highest values in ponds 4. No significant differences in vertical profiles were observed nor were there significant variations between morning and afternoon.

Oxygen concentrations increased slowly along the system. Significant differences were found between the surface layer and the middle layer for all ponds. Only the last three ponds reached concentrations higher than 1 mg l^{-1} in the upper layer. Figure 2a presents average oxygen concentrations for morning and afternoon measurements.

Temperature, pH and oxygen profiles. Phase II
No longitudinal temperature gradients were observed along the system. Vertically, morning temperatures ranged between 22 and 24 °C, with significant difference between the surface and middle layers. In the afternoon, the vertical gradient was 24 - 27.5 °C, with significant differences between surface, middle and bottom layers.

The pH range was 6.7 – 7.5, the highest value being observed in pond 3. Vertical stratification was present only in pond 1 with significant difference between middle and bottom layers with the lowest value in the bottom.

Oxygen concentrations were found to be significantly different between surface and middle layer, with exception of pond 1 and 3. For ponds 1 and 3 differences were not significant due to the large variations observed in these two ponds (Fig. 2).

Oxygen concentrations in Phase II were significantly higher in all ponds (comparing same layer and same pond number) than in phase I. In the surface layer, oxygen was higher than 1 mg l^{-1} for all ponds and close to the bottom layer oxygen was higher than 0.7 mg l^{-1} after pond 3.

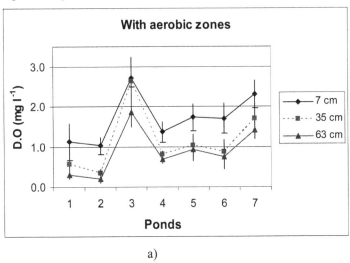

a)

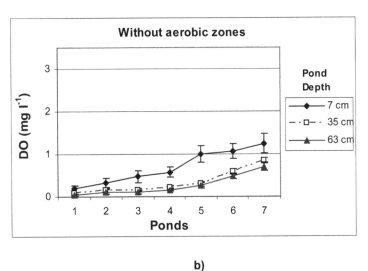

b)

Fig. 2. Dissolved oxygen profiles during a) Phase I and b) Phase II

UASB effluent characteristics
The characteristics of the influent to the systems (effluent of UASB reactor) during this study are presented in Table 1.

Nitrogen removal patterns
Ammonium nitrogen concentration was significantly lower during the second phase than during the first one, in all ponds (Fig.3). Final ammonium nitrogen concentrations were 5.9 ± 2.1 mg NH_4^+-N l^{-1} (81% removal) and 0.6 ± 0.6 mg NH_4^+-N l^{-1} (98% removal) for first and second phase, respectively. Total Kjeldahl nitrogen concentrations were also significantly different between corresponding pond numbers of phase I and II. Final effluent concentrations of 10.5 ± 3 mg TKN l^{-1} (71% removal) and 2.6 ± 0.8 mg TKN l^{-1} (93% removal) were observed for first and second phase, respectively.

Table 1 UASB reactor effluent characteristics

PARAMETER	UNITS	UASB EFFLUENT*
pH	-	6.5 -7.2
Temperature	°C	24.6 ± 1.3
COD	mg l^{-1}	189 ± 49
BOD$_5$	mg l^{-1}	111 ± 23
TKN	mg l^{-1}	36.8 ± 5.5
N-Ammonium	mg l^{-1}	30.5 ± 5.3
N-Organic	mg l^{-1}	6.3 ± 2.2
N-Nitrites	mg l^{-1}	0.02 ± 0.01
N-Nitrates	mg l^{-1}	0.02 ± 0.01
Total Phosphorus	mg l^{-1}	6.8 ± 2.0
Total Solids	mg l^{-1}	359 ± 24
Total Suspended Solids	mg l^{-1}	57 ± 15
Conductivity	$\mu S\ cm^{-1}$	624 ± 31

Averages ± S.D. (n=12 daily average; each n = average of 4 or 6 integrated samples/day)

During the first phase, oxidized nitrogen concentrations in the effluent were very low until pond 4 and increased from pond 5 to pond 7 with final concentration of 3.5 ± 0.8 mg NOx-N l^{-1}. During the second phase, low nitrate concentrations were observed from the second pond onwards with a final concentration of 1.5 ± 0.5 mg NOx-N l^{-1}. Final oxidized nitrogen concentrations were significantly different between the two phases.

Effluent total nitrogen was significantly different between the two phases, with 13.8 ± 2.9 mg TN l^{-1} (63 % removal) and 3.7 ± 1.5 mg TN l^{-1} (90% removal) for first and second phase, respectively. Statistical analysis showed significant differences between the two phases for comparison between corresponding pond numbers, with exception of ponds 1. Within each phase, effluent total nitrogen concentrations were significantly different between consecutive ponds.

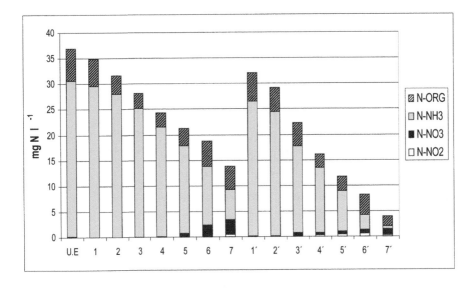

Fig. 3. Nitrogen removal patterns in the effluent of individual ponds during phases I (ponds 1 to 7) and II (ponds 1' to 7'). UE stands for UASB effluent

Nitrogen Balance
The nitrogen balances established per pond during each phase are presented in Figure 4. Detailed information on the methodology is presented elsewhere (Caicedo *et al.*, 2004). Influent and effluent volumetric flows were less than 1% different, which was considered negligible. The term N_f corresponding to nitrogen fixation is assumed negligible because the presence of ammonia represses nitrogen fixation (Brock *et al.*, 1991; Duong and Tiedje, 1985). Figure 5 presents the overall nitrogen balances over the system during the two phases. Overall nitrogen removal rates were 765 mg N m^{-2} d^{-1} in phase I and 1089 mg N m^{-2} d^{-1} in phase II.

Discussion
The introduction of aerobic zones did have a significant effect on the level of oxygen in the system as can be observed in Figure 2. During the first phase, the oxygen concentrations were under 0.5 mg l^{-1} up to pond 4 and only from the fifth pond on oxygen concentrations in the surface layer reached higher concentrations. This was reflected in higher nitrification rates towards the end of the system. An oxygen concentration of O.5 mg O_2 l^{-1} is the lower limit for nitrification to occur, as reported by Eighmy and Bishop (1989) and Metcalf and Eddy (2003). During the second phase, oxygen concentrations were above this limit in the upper layer for all ponds and in the middle and bottom layer from pond 2 onwards.

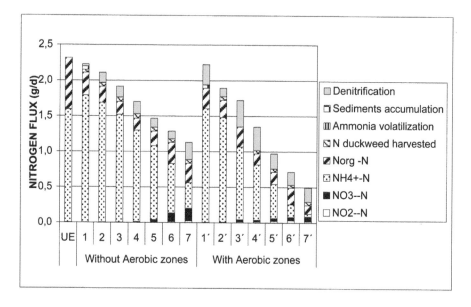

Fig. 4. Nitrogen balances per pond during phase I (left) and phase II (right). UE stands for UASB effluent.

The contribution of ammonia volatilization to nitrogen removal was very low in both phases. This finding is in agreement with Zimmo's results (2003b) and with Ferrera and Advi's conclusions (1982). Nitrogen removal by sedimentation was also low and not significantly different in both phases. Nitrogen biomass uptake was the second most important nitrogen removal mechanism in both phases. Although biomass production per square meter in covered areas was not significantly different the nitrogen removal was lower in the second phase due to the lower area available for growth.

Denitrification was the most important removal mechanism in both phases (Fig. 5) and was clearly higher during the second phase. Denitrification rates were in the range of 112-937 mg N m^{-2} d^{-1} and 446-1414 mg N m^{-2} d^{-1} during phases I and II, respectively. These ranges are higher than those reported by Zimmo et al. (2003a) for a duckweed pond system (265 – 409 mg N m^{-2} d^{-1}) under similar environmental conditions and higher than the value reported by Vermaat and Hanif (1998) in a laboratory scale experiment (260 mg N m^{-2} d^{-1}). As reported earlier (Caicedo et al., 2004), the compartmentalized configuration of the system plus the up and down flow pattern may explain the higher ranges of denitrification rates found in this study. In phase II, the denitrification rate reached even higher values than in phase I. This may be explained by the presence of the aerobic zones, which affect the overall amount of nitrogen being nitrified and subsequently denitrified (Fig. 5). The aerobic zones also changed the profiles of nitrification and denitrification rates along the systems (Fig. 6).

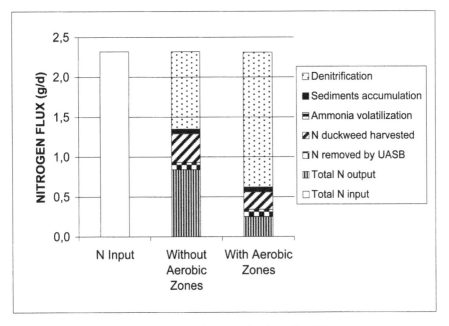

Fig. 5. Fate of Nitrogen in duckweed pond systems in phases I and II.

During phase I, the nitrification rate was low in the first pond, where the average oxygen concentration was low. It was fairly constant between ponds 2 and 5 where a small increase in oxygen concentration was observed, especially in the surface layer. In pond 6 and 7, oxygen concentration increased above 0.5 mg l^{-1} even in the bottom layers and this was reflected in higher nitrification rates and accumulation of nitrates in these last two ponds (Fig. 3). No limitation by ammonium nitrogen concentration was expected as the NH_4^+-N was above 5.9 mg N l^{-1} in all ponds, a value considerably higher than the half saturation coefficient (1 mg NH_4^+-N) suggested by Henze et al. (1986, as cited by Water Environmental Federation, 1998).

During phase II, the pattern of nitrification rates along the system was very different from phase I. In ponds 1 and 3 (the uncovered ponds), nitrification rates were 400% and 250% higher than in the same ponds in phase I. In pond 4 (a covered pond) nitrification rate was also higher (78%) than pond 4 in phase I, despite the decrease in oxygen concentrations. The explanation for this may be that part of the oxidized nitrogen that had been produced in pond 3 was transferred from the top layer of pond 3 to the bottom layer of pond 4 to be denitrified and was accounted for in the nitrification rate calculation of pond 4. This was not the case in pond 2 possibly due to lower oxidized nitrogen production in pond 1. Pond 7 in phase II presented lower nitrification rates than the same pond in phase I in spite of higher oxygen concentration perhaps due to the low ammonium nitrogen concentration present in this pond.

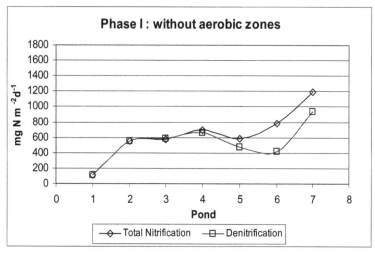

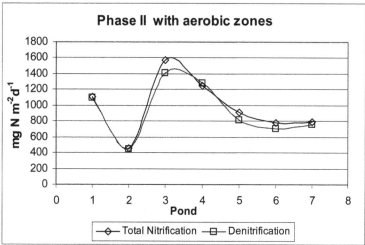

Fig. 6. Nitrification and denitrification rates during Phase I and II. 'Total nitrification' refers to denitrification plus Nox measured in the water phase.

Nitrification has been reported to be dependant on two limiting factors, oxygen and ammonium nitrogen (Water Environmental Federation, 1998; Metcalf and Eddy, 2003). A multiple linear regression with stepwise selection (p=0.05), performed between nitrification rate vs. oxygen and ammonium nitrogen concentrations, eliminated ammonium nitrogen as a variable. With the data of the two phases, excluding pond 7 in phase II where ammonium nitrogen may have been limiting, two regression models were tested, one with the average oxygen concentration in the ponds (upper, middle and bottom) and the other with the oxygen concentrations in the upper layer. The last one produced better results (Eq. 2; Fig. 7).

$$r_N = 658.9 \times O_2 \quad (R^2 \text{ adjusted} = 0.89) \qquad (2)$$

where:

r_N :   Nitrification rate (mg N m^{-2} d^{-1})
O_2:   Oxygen concentration (mg O_2 l^{-1})

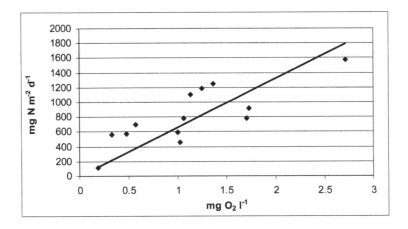

Fig 7. Plot of nitrification rate vs. oxygen concentration in the upper layer during phases I and II.

The positive linear relationship between the nitrification rates vs. oxygen concentration shown in this study is in disagreement with the saturation type model commonly proposed in the literature (Water Environmental Federation, 1998; Metcalf and Eddy, 2003; Zimmo et al., 2003a). Reported values of oxygen half saturation constant for oxidation of ammonia (the limiting step of nitrification) ranges from 0.3 to 0.5 mg O_2 l^{-1} (Water Environmental Federation, 1998). Figure 7 shows that nitrification rate continued to increase at oxygen concentrations higher than this range. It has to be taken into account that the Monod saturation model for biomass growth or the Michaelis-Menten model for reaction velocity (Bailey and Ollis, 1986) were defined for completely mixed reactors where the concentrations of substrates and biomass are equal in all points of the reactor. The ponds of the experimental system of this study were not completely mixed reactors. Figure 2 shows not only clear oxygen stratification from surface to bottom, but also a higher aerobic zone as the surface oxygen concentration increased which means that a higher percentage of the pond volume will be aerobic and as a consequence it will produce more oxidized nitrogen per unit area. From the above it can be concluded that the application of the half saturation kinetics needs to duckweed stabilization ponds need to be subject of further research.

It can be observed that the denitrification rate follows the pattern of the nitrification rate (Fig. 6), which suggests that denitrification rate was controlled by nitrification. The occurrence of denitrification in the aerobic environment present in some of the ponds indicates the existence of localized anoxic places in the biofilms of the system (Christensen and Harremoes, 1978).

The system with aerobic zones achieved higher average nitrogen removal rates 1089 mg N m^{-2} d^{-1} than the system without aerobic zones 765 mg N m^{-2} d^{-1}. From previous discussion, it can be concluded that this higher removal is mainly due to the increase of nitrification-denitrification. Zimmo *et al.* (2003c) in an experiment with a similar experimental set up and temperature, found higher removal rate in a conventional stabilization pond system (1448 mg N m^{-2} d^{-1}) than in a duckweed covered stabilization pond system (1308 mg N m^{-2} d^{-1}). Conventional stabilization ponds may offer the possibility of higher nitrogen removal but the nitrification process may be inhibited if high pH fluctuations occur. Duckweed systems may yield lower nitrification-denitrification but additional nitrogen removal is achieved by plant uptake and biomass harvesting. A combination of conventional and duckweed stabilization ponds may offer the possibility of obtaining good nitrogen removals and at the same time to recover part of it as protein biomass. The effluent of the system with aerobic zones fulfilled the European Standard for effluent nitrogen in wastewater treatment plants (< 10 mg l^{-1}) at pond 6 (HRT=18 d) and ammonium nitrogen was already lower than this value at pond 5 (HRT= 15 d). The system without aerobic zones would have achieved the EU standard at retention time higher than 21 days.

In phase 1, nitrification occurred at low oxygen concentrations, although not at a high rate, but still enough to make denitrification an important part of the nitrogen balance. In phase II, nitrogen transformations and overall nitrogen balance were changed significantly, as well as the effluent nitrogen concentration. The results confirmed the hypothesis that the introduction of aerobic zones in early stages of the system could enhance the nitrification-denitrification process in the entire system. Similar observations were made when aerobic unplanted zones were introduced in a constructed wetland (Okia, 2000). The results of this study demonstrate the potential of inclusion of aerobic zones in a duckweed system to enhance nitrification and consequently denitrification rates.

Conclusions
- Ammonia volatilization and sedimentation were insignificant processes for nitrogen removal in a series of duckweed ponds fed with anaerobically pre-treated municipal sewage.

- Nitrification played an important role in nitrogen transformations in the duckweed systems and it was affected by the introduction of aerobic zones.

- Denitrification also plays a key role in nitrogen transformations and removal. Despite the presence of oxygen in the water column, denitrification probably occurred in the microenvironment of the biofilms on the pond-walls, on the plant biomass and in the bottom sludge. Denitrification was the most important removal mechanism during the two experimental phases of the present research.

- The inclusion of aerobic zones in early stages of the system increases significantly the nitrification-denitrification rates in duckweed systems.

- Higher nitrogen removals may be obtained in duckweed pond systems through the introduction of aerobic zones, which allows a considerable reduction of the hydraulic retention time. Strict nitrogen effluent criteria can therefore be met at relatively short hydraulic retention times.

Acknowledgements
The authors would like to thanks Mr. M. Marquez for his help in the collection of the data, to Ms I. Yoshioka for her collaboration in the statistical analysis, to the Netherlands's Government and Universidad del Valle for their support in the development of this research through the SAIL ESEE project.

References
A. P. H. A. (1995). American standard methods for the examination of water and wastewater. 19th edition. New York.

Bailey J. E. and Ollis D. F. (1986). Biochemical Engineering Fundamentals. 2nd Edition. Mc Graw Hill, New York.

Caicedo J. R., Steen N. P., Gijzen H. J. (2003). The effect o anaerobic pre-treatment on the performance of duckweed stabilization ponds. Proceedings of International Seminar on natural systems for wastewater treatment. Agua 2003. Cartagena Colombia.

Caicedo J. R., Steen N. P. van der , Gijzen H.J. (2004). Nitrogen balance of duckweed covered sewage stabilization ponds. In this thesis.

Christensen M. H., Harremoes P. (1978). Nitrification in wastewater treatment. In Mitchell R. (Ed.) Water Pollution Microbiology. Vol. 2. John Wiley & Sons.Inc. N. Y. 391-414.

Daniel W. W. (1990). Applied nonparametric statistics. Georgia State University. Houghton Mifflin Company. Boston.

Grace S. A. (1998). Optimisation of nitrogen removal in an afro-tropical constructed wetland. MASTER THESIS. IHE.Delft – Holland.

Garcia F. (1999). Nitrogen removal in a cyperus papyrus constructed wetland with alternating planted and cleared zones. Master Thesis. IHE, Delft-Holland.
Gijzen H. J. (2001). Anaerobes, aerobes, and phototrophs: a wining team for wastewater management. *Wat. Sci Tech.*, 44(8), 123-132.

Metcalf and Eddy. (1991). Waste Engineering. Treatment, disposal and reuse. Tchobanoglous G. and Burton F. L. [eds.]. 2nd Ed. McGraw Hill, Inc. USA.

Okia T. (2000). A pilot study on municipal wastewater treatment using a constructed wetland in Uganda. Ph. D. Dissertation. Wageningen University and Internationtal Institute for Infrastructural, Hydraulic and Environmental Engineering, Holland.

Oostrom A. J. van (1995). Nitrogen removal in constructed wetlands treating nitrified meat processing effluent. *Wat. Sci. Tech.*, 32 (3), 137-147

Oron G., Porath D., Wildschut L.R. (1986). Wastewater treatment and renovation by different duckweed species. *J. Environmental Engineering Division, ASCE*, 112(2), 247-263.

Oron G., Porath D., Jansen H. (1987). Performance of the duckweed species *Lemna gibba* on municipal wastewater for effluent renovation and protein production. *Biotech. & Bioeng.* 29(2), 258-268.

Shilton A., Mara D. D., Pearson H. W. (1995). Ammonia volatilisation from a piggery pond. Wat. Sci. Tech., 33 (7), 183-189.

Skillicorn P., Spira W., Journey W. (1993). Duckweed aquaculture, a new aquatic farming system for developing countries. The World Bank. 76 p. Washington.

Vermaat J. E., Hanif M. K. (1998). Performance of common duckweed species (*Lemnaceae*) and the waterfern (*Azolla filiculiodes*) on different types of wastewater. *Wat. Res.*, 32 (9), 2569-2576.

Water Environment Federation (1998). Biological and chemical systems for nutrient removal. Alexandria, USA.

White K. D. (1995). Enhancement of nitrogen removal in sub-surface flow constructed wetlands employing a 2-stage configuration, an unsaturated zone, and re-circulation. Wat.]Sci. Tech. Vol. 32 (3), 59-67.

Zimmo O., Al-Sa'ed R. M., Van der Steen N. P, Gijzen H. J. (2002). Process Performance Assessment of algae-based and duckweed-based wastewater treatment systems. *Wat. Sci. Tech.*, 45(1), 91-101.

Zimmo O., Van der Steen N. P, Gijzen H. J. (2003a). Quantification of nitrification and denitrification rates in pilot-scale algae and duckweed-based waste stabilization ponds. In: Nitrogen transformation and removal mechanism in algal and duckweed waste stabilizacion ponds. Ph. D. Dissertation. Wageningen University and International Institute of Hydraulic and Environmental Engineering. Holland.
Zimmo O., Steen N. P. van der, Gijzen H. J. (2003b). Comparison of ammonia volatilisation rates in algae and duckweed-based waste stabilization ponds treating domestic wastewater. *Wat. Res.* 37, 45587-4594

Zimmo O., Steen N. P. van der, Gijzen H. J. (2003c). Nitrogen mass balance over pilot scale algae and duckweed-based wastewater stabilization ponds. In: Nitrogen transformation and removal mechanism in algal and duckweed waste stabilization ponds. Ph. D. Dissertation. Wageningen University and International Institute of Hydraulic and Environmental Engineering. Holland.

Chapter 7

Effect of pond depth on removal of nitrogen in duckweed stabilization ponds

Chapter 7

Effect of pond depth on removal of nitrogen in duckweed stabilization ponds

Abstract.
The effect of pond depth on nitrogen removal in duckweed stabilization ponds was studied in a pilot plant consisting of two lines with seven duckweed ponds in series, with different depths and fed with effluent of a laboratory scale UASB reactor. Three experimental conditions were studied: 1) Pond depth 0.7 m, HRT= 21 days, 2) Pond depth 0.4 m, HRT = 12 days, 3) Pond depth 0.4 m, HRT = 21 days. The systems were evaluated based on pH, temperature and oxygen profiles, organic matter removal (BOD_5), nitrogen transformations, biomass production and biomass nitrogen content.

Average total nitrogen removal rates were 598 mg N m^{-2} d^{-1} for DSP 1, 589 mg Nm^2 d^{-1} for DSP 2 and 482 mg N m^{-2} d^{-1} for DSP 3. In spite of the lower nitrogen removal rate in DSP 3, it has the higher removal efficiency (44 %, 43 % and 62 % for DSP 1, 2 and 3 respectively) due to the lower surface loading rate in this system. This shows that using the percentage of removal as a parameter for comparison should be done with care and the operational parameters of the compared systems should be taken into account. Denitrification was the most important nitrogen removal mechanism for the three DSPs. Nitrogen removal via biomass production was the second most important removal mechanism for the three experiments. Pond depth does not seem to determine nitrification or denitrification,. Nitrification seems to be related to surface organic loading rate, while denitrification was related to BOD availability. The comparison between two pond systems with different depths, but operated at the same hydraulic surface loading rate (DSP 1 and 2) showed similar nitrogen removals in the shallower system as in the deeper system. This suggests that duckweed pond system could be designed with the shallower depth without affecting surface loading and nitrogen removal efficiency. Nitrogen removal appeared to be governed by the surface loading rate rather than by the hydraulic retention time.

Introduction
Duckweed stabilization ponds are natural systems for wastewater treatment. This technology has gained interest because of the contribution of the duckweed plants in nutrient recovery and re-use. Duckweed shows rapid growth on sewage, and has high protein content. Besides, DSPs have been shown to be a low cost alternative with easy operation and maintenance (Gijzen and Ikramullah, 1999).

The design of conventional algal based stabilization ponds is based on the organic surface-loading rate and assumes no relation to pond depth (Metcalf and Eddy, 1991; Mara *et al.*, 1992). Nitrogen removal in these systems has not been studied fully nor has it been optimised. DSPs, being a more recent technology, need further research in order to define design criteria for optimal performance in the removal of different treatment parameters.

Craggs *et al.*, (2002) working with an advanced pond system consisting of an advanced facultative pond, a high rate pond, an algae pond and a maturation pond, did not find noticeable differences in pond performance when comparing two depths for the high rate ponds. Silva *et al.* (1995) proposed shallow maturation ponds to enhance N removal via ammonia volatilization. Zimmo (2003a), who has developed a model for nitrogen removal in conventional stabilization ponds, came to the conclusion that pond depth plays an important role in nitrogen removal. Caicedo *et al.* (2004a) studied N-removal in DSP and speculated on ways to improve nitrogen removal. Pond depth might affect oxygen balances in the system and as a consequence nitrogen transformation and area related process like growth of nitrifiers. The paper describes the results of a study on the effect of pond depth on nitrogen removal in a duckweed system.

Materials and methods

Experimental set up
The experiment was carried out in the Universidad del Valle, located in the city of Cali in Colombia. The climatic conditions of the region are tropical, altitude is 1000 meters above sea level, latitude is 3°27 ' and longitude 76°31 '.

The laboratory scale UASB reactor used in the experiments had a volume of 23 litres and was operated at a hydraulic retention time of 6 hours. The duckweed systems consisted of seven duckweed ponds in series, with a water volume of 165 litres each for DSP 1 and 94 litres each for DSP 2 and 3 (Fig. 1).

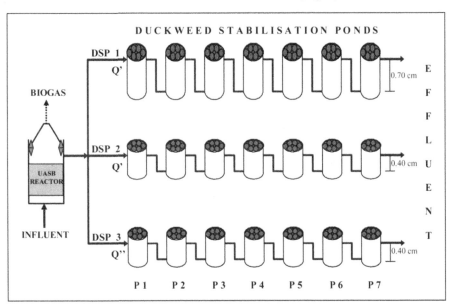

Fig. 1 Diagram of the set-up for DSP 1, 2, 3. Series of ponds: P1 to P7. Q'= flow for DSP 1 and 2, Q''= flow for DSP 3. Pond depth: DSP 1= 0.70 cm., DSP 2 and 3 = 0.40 cm.

Artificial wastewater.
The set-up was fed with effluent of a pilot scale UASB reactor, which was receiving synthetic wastewater with characteristics similar to that of domestic wastewater (protein 50%, starch 24%, cellulose 8%, oil and detergent 10%, percentages in COD basis). The wastewater was prepared in tap water and contained 418 ± 27 mg COD l^{-1}, 278 ± 35 mg BOD_5 l^{-1}, 44.8 ± 2.9 mg TKN l^{-1} and 4.1 ± 0.8 mg TP l^{-1}. In addition, macro and micronutrients were added as indicated in Table 1.

Table 1. Macro and micronutrient concentrations in the artificial wastewater

Macro nutrients	mg l^{-1}
Urea	42.9
K_2HPO_4	11.9
KH_2PO_4	8.8
$MgCl_2.6H_2O$	7.0
NaCl	40.0
Micro nutrients	mg l^{-1}
EDTA	13.3
$FeCl_3.6H_2O$	4.4
$MnSO_4.H_2O$	0.09
$CoSO_4.7H_2O$	0.03
$ZnSO_4.7H_2O$	0.03
H_3BO_3	0.01
$(NH_4)8Mo_7.O_{24}.4H_2O$	0.02
$Na_2SeO_3.5H_2O$	0.03
$NiCl_2.6H_2O$	0.12
$CuSO_4$	0.04

Operational conditions.
Three DSP were operated in parallel. The operational parameters for each treatment are shown in Table 2. DSP 1 and 2 were started simultaneously and operated until reaching steady state, before starting the three months sampling program. DSP-3 was started after finalising the work on DSP-2. This meant that the influent flow rate was reduced. Three weeks of stabilization was allowed before starting the sampling period. DSP 1 was continued during the testing of DSP 3.

Table 2. Operational parameters for DSP 1, 2, 3.

Parameter	Units	DSP 1	DSP 2	DSP 3
Flow	$l\,d^{-1}$	55	55	31.4
Depth	m	0.70	0.40	0.40
HRT per pond	d	3	1.7	3
Total HRT	d	21	12	21
Hydraulic surface loading	$m^3\,m^{-2}\,d^{-1}$	0.23	0.23	0.13
Organic loading rate to the first pond	Kg BOD_5 $ha^{-1}d^{-1}$	184	184	105
Biomass Density (after harvesting)	g fresh weight m^{-2}	700	700	700

Due to the constant composition of the wastewater, grab samples were considered suitable to characterize the influent to the systems. Once the systems reached steady state, once a week grab samples were also taken at the outlet of each pond. pH, temperature and oxygen profiles were measured during the morning every two weeks.

Biomass harvesting, sampling and analysis
The species of duckweed used was *Spirodela polyrrhiza*, which was collected in the surroundings of the area of study. This species is available in the Valle region and it has shown good adaptation to domestic wastewater.

Every 4 days part of the duckweed cover was harvested. Biomass samples were taken with a strainer of a known area and it was allowed to drain for about 5 minutes. The fresh weight was determined and the density was calculated. Enough biomass was harvested to leave a density of 700 g m^{-2} (fresh weight) at the beginning of each harvesting period. Previous experiments showed that this density prevents light penetration and algae growth. The values for biomass production of every two harvesting periods were grouped and production was calculated in g d^{-1} for each pond.

Every two weeks, the total of harvested biomass from each pond was mixed and 20 grams of fresh weight (FW) were dried at 70°C for 24 hours to determine the dry weight. The nitrogen content was determined in the dried biomass.

Water samples analysis.
Biochemical oxygen demand (BOD$_5$), chemical oxygen demand (COD), Total Kjeldahl Nitrogen (TKN), Ammonium Nitrogen (NH$_4^+$-N), Nitrate Nitrogen (NO$_3^-$-N) and Nitrite Nitrogen (NO$_2^-$-N) of influent and effluent samples were measured according to the American Standards Methods (APHA, 1995). Dissolved oxygen, pH, temperature were measured with portable electrodes directly in the ponds.

Nitrogen Balance
The law of conservation of matter was used as the basis for mass balance calculations. The mass balance equation that was used is the following:

$$N_{in} = N_{out} + N_{dw} + N_d + N_v + N_s - N_f \qquad (1)$$

Where,

N_{in}	= (N_{kj} + NO$_3^-$ + NO$_2^-$) in the influent
N_{out}	= (N_{kj} + NO$_3^-$ + NO$_2^-$) in the effluent
N_{kj}	= Kjeldahl nitrogen (Organic nitrogen + ammonia nitrogen)
N_{dw}	= N removed via duckweed harvesting
N_d	= N removed via denitrification
N_v	= N removed via ammonia volatilisation
N_s	= N-accumulation in sediment
N_f	= N-fixation

All fluxes are expressed in g d⁻¹. Influent and effluent nitrogen loads were obtained by multiplying the flow with the total nitrogen concentration [NH_4^+ + N Org + NO_2^- + NO_3^-]. Nitrogen biomass uptake flux is determined by multiplying biomass production per % of dry solids and per % of nitrogen content of the biomass. Previous experiments showed less than 1% of nitrogen volatilization at slightly higher pH than prevalent in the current experiments (Caicedo *et al.*, 2004a). The nitrogen flux through volatilization is therefore assumed negligible in these experiments. The accumulation of nitrogen in sediments was negligible due to none presence of suspended solids in artificial wastewater, low algae growth and low accumulation of dead biomass in the bottom of the ponds. This is agreement with previous studies in similar experimental set up (Caicedo *et al.*, 2004a). Nitrogen fixation is assumed negligible because it is suppressed by presence of ammonium nitrogen (Brock *et al.*, 1991). Denitrification was calculated as the difference between the nitrogen input and all other nitrogen fluxes from the system. Total nitrification was calculated by adding the denitrification flux plus the oxidized nitrogen flux leaving the system in the effluent. Net nitrification was calculated as the flow of oxidized nitrogen leaving the systems.

Data analysis
The SPSS statistical package was used to analyse the results. The comparisons when needed were performed with the non-parametric Krusk and Wallis method (Daniel, 1990) at 95% level of confidence.

Results

UASB effluent composition and organic matter removal
The UASB reactor effluent showed a stable composition in terms of pH, organic matter, and nitrogen compounds. The results are presented in Table 3.

Table 3. UASB effluent characteristics.

Parameter	Concentrations mg l⁻¹
COD	99 ± 22 (n=20)
BOD$_5$	79 ± 8 (n=15)
NH_4^+-N	40.2 ± 3.5 (n=20)
TKN	42.8 ± 2.4 (n=20)
NO_2-N	0.01 ± 0.01(n=20)
NO_3-N	0.05 ± 0.05 (n=15)
Total Phosphorus	4.2 ± 0.7 (n=23)

The UASB reactor was very efficient in removing organic matter (72 % BOD$_5$ removal). As a consequence the pond systems were working at low organic loading rate, with values between first and last pond of 184-32, 184-29, 105-19 Kg BOD$_5$ha⁻¹ d⁻¹ for DSP1, 2 and 3 respectively. The patterns of BOD$_5$ removal along the systems for the three DSPs are presented in Figure 2.

Environmental conditions.
Water temperature ranged between 25.5 - 24 °C from surface to bottom for DSP 1 and 2 and between 24.5 - 23 °C for DSP 3. The measurements in DPS 3 were taken

an hour earlier than in DPS 1 and 2. pH started near neutrality in the first pond for the three DSPs and decreased progressively along the ponds. For DSP 1 and 2 the pH ranged between 7 (pond 1) to 6 (pond 7), while in DSP 3 pH ranged between 7 (pond 1) down to 4.5 in the last pond.

Dissolved oxygen concentrations increased gradually along the seven ponds for the three DSPs (Fig 3). Oxygen concentrations reached 1.1, 1.1 and 3.9 mg l^{-1} in the last pond for DSP 1, 2 and 3 respectively.

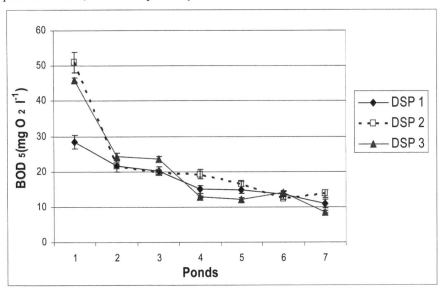

Fig. 2. Effluent BOD_5 concentration for DSP 1, 2 and 3. Error bar indicates standard error (n=7).

Biomass production
For the three DSPs, biomass production was higher in the first pond and reduced gradually until the last pond (Fig 4). Biomass production in the last two ponds of DSP 3 was significantly different (p=0.05) from the biomass production in the same ponds in DSP 1 and 2.

Effluent composition in terms of the different nitrogen compounds.
The profiles of the different nitrogen compounds (ammonium, organic nitrogen, nitrites and nitrates) along the systems for the three DSPs, is presented in the Figure 5. For comparison, nitrogen present in the influent is also shown. Effluent nitrogen concentrations were not significantly different between DSP 1 and 2 and significantly different between DSP 1 and 3.

Nitrogen Balance
Previous experiments have demonstrated the importance of nitrogen balances to get a better understanding of the nitrogen transformation processes occurring in duckweed systems.

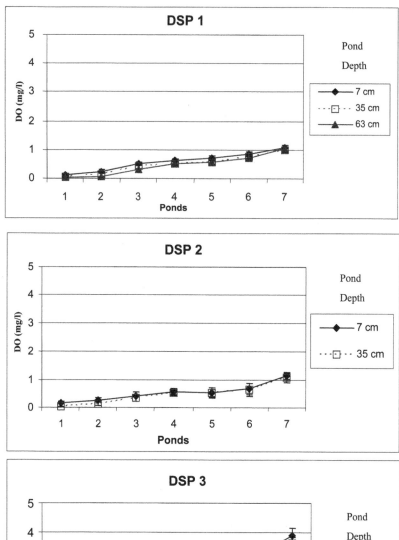

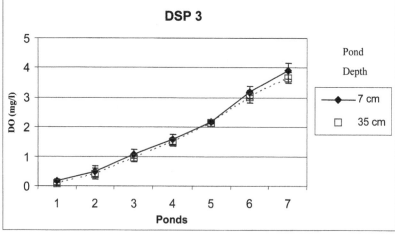

Fig 3. Oxygen profiles for DSP 1, 2 and 3. Error bar indicates standard error. (n=8 for DSP 1, n=5 for DSP 2 and 3).

Influent and effluent flows were measured regularly and differed less than 1% of average daily flow. No precipitation was taken into consideration as the systems were protected from the rain. The results for nitrogen biomass uptake flux are presented in Table 4. The results of the mass balances per pond are shown in Figure 6. Percentages of removed nitrogen load are presented in Table 5.

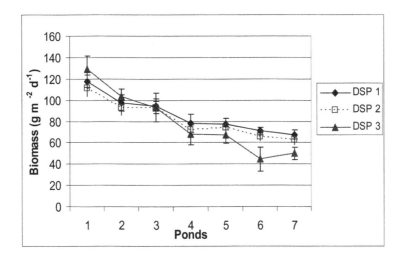

Fig 4. Biomass production for DSP 1, 2 and 3. Error bar indicates standard error.

Table 4. Nitrogen biomass up-take for DSP 1, 2 and 3

Parameter	Units	DSP 1	DSP 2	DSP 3
Biomass production (fresh weight)	g d^{-1}	142 ± 23 (n=24)	135 ± 19 (n=12)	131 ± 26 (n=12)
Dry solids	%	4.5 ± 0.1 (n=10)	4.6 ± 0.1 (n=7)	4.2 ± 0.1 (n=7)
Nitrogen content of dry solids	%	5.5± 0.1 (n=10)	5.6 ± 0.1 (n=7)	5.2 ± 0.1 (n=7)
Nitrogen biomass up-take	g d^{-1}	0.36	0.35	0.29

Table 5. Percentage of removed nitrogen for the different removal mechanisms for DSP 1, 2 and 3.

Description	DSP 1 %	DSP 2 %	DSP 3 %
Total N input	100	100	100
Total N output	56	57	38
N duckweed harvested	15	15	22
Ammonia volatilization	0	0	0
Sediments accumulation	0	0	0
Denitrification	29	28	40

Effect of Operational Variables on Nitrogen Transformations in Duckweed Stabilization Ponds

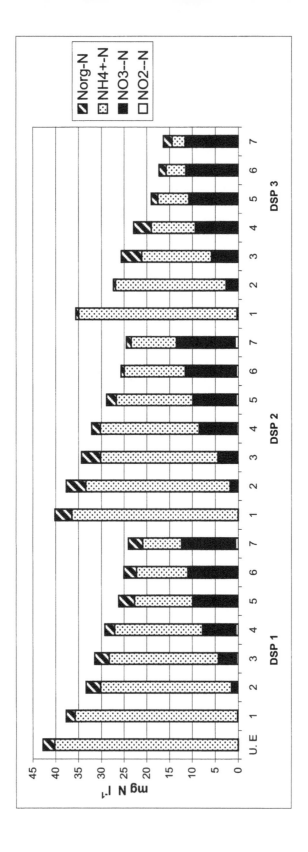

Fig 5. Influent and effluent concentration of different nitrogen compounds in DSP 1, 2 and 3.

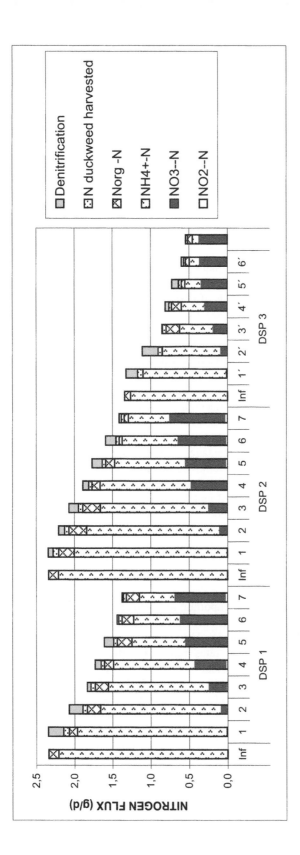

Fig. 6. Nitrogen balances per pond in DSP 1, 2 and 3.

121

Discussion

From Table 5 it can be seen that total nitrogen removals in terms of influent loads were ⸱ %, 43 % and 62 % for DSP 1, 2 and 3 respectively. DSP 3 seems more efficient in terms percentage, but from the nitrogen balance it can be seen that the actual amount of nitrog removed is lower than in the other two DSPs. Average total nitrogen removal rates we 598 mg N m^{-2} d^{-1} for DSP 1, 589 mg N m^{-2} d^{-1} for DSP 2 and 482 mg N m^{-2} d^{-1} for DSP This apparent inconsistency may be explained by the difference in loading rate betwe DSP 1 and 2, and DSP 3. This is showing that using a percentage of removal as parame should be done with care, and the operational parameters of the compared systems shou be taking into account.

The diminishing of organic matter and nutrients along the systems may explain t reduction of biomass production along the systems. The significant lower bioma production in the last two ponds of DSP 3 compared to the production of the same ponds DSP 1 and 2 is probably caused by the low pH present in these ponds (pH = 4.5 - : *Spirodela polyrrhiza* is reported to grow well in the pH range of 6 – 7.5 with an optimu grow at pH = 7 (Landolt and Kandeler, 1987; Caicedo et al., 2000). The low pH was caus most probably because of nitrification which gradually consumed the available alkalinity the effluent of the UASB reactor.

Similar to previous reports (Caicedo et al., 2004a; Caicedo et al., 2004b) nitrogen remov via biomass production was the second removal mechanism (15 %, 15 % and 22 % for D! 1, 2 and 3 respectively). Values for DSP 1 and 2 are the same. Since the surface loadi rates are the same, these percentages represent similar amounts of nitrogen removed. DSP 3 the amount of nitrogen removed was lower but it represents higher percentage of t influent load. Biomass productions were in the range of 113–65, 106–60, and 124–48 g n d^{-1} (fresh weight) for DSP 1, 2 and 3 respectively. These values are in the same range reported in earlier work (Caicedo et. al., 2003). Nitrogen biomass up-take was in the ran of 312-161, 305-155, and 302-105 mg N m^{-2} d^{-1} for DSP 1, 2 and 3 respectively. Also the values are similar to our previous results (Caicedo et al., 2004a), and to results reported literature for *Spirodela polyrrhiza* (Kvet et al., 1979; Reddy and Debusk, 1985; Alaerts al., 1996).

Denitrification was the most important nitrogen removal mechanism; rates from first to la pond were in the range of 820 – 70 mg N m^{-2} d^{-1} (DSP 1), 547-115 mg N m^{-2} d^{-1} (DSP : and 681-19 mg N m^{-2} d^{-1}(DSP 3). The average values 382, 371, and 306 mg N m^{-2} d^{-1}, we considerably lower than the rates reported earlier (Caicedo et al., 2004a) in simil experimental systems, treating real domestic wastewater. Values were comparable with t ones reported by Zimmo (2004) in a pilot scale duckweed pond system. Assuming that 3 mg BOD$_5$ l^{-1} are needed to denitrify 1 mg NO$_3$ l^{-1} (Oostrom, 1995; Caicedo et al., 2004 the amount of BOD$_5$ required for full denitrification of remaining nitrate in the ponds w calculated. These calculations show that, except for the first two ponds of the systems, other ponds showed insufficient BOD$_5$ availability. This explains the decreasing tenden of the denitrification rates along the systems (Fig. 7).

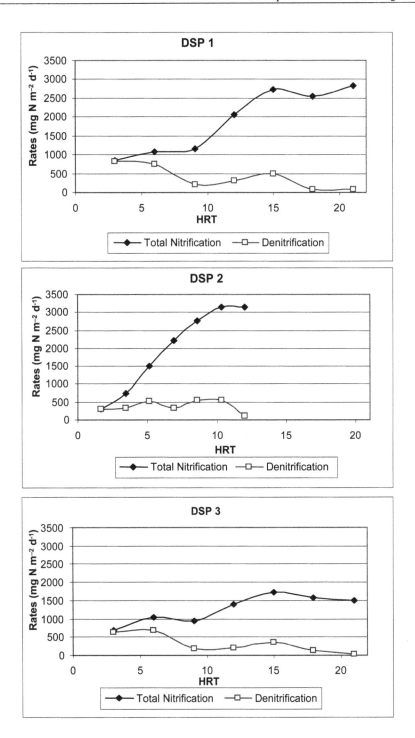

Fig 7. Total nitrification and denitrification rates in DSP 1, 2 and 3

Total nitrification rates were in the range of 853 –2834 mg N m^{-2} d^{-1} (DSP 1), 11 –3145 m N m^{-2} d^{-1} (DSP 2) and 682 – 1730 mg N m^{-2} d^{-1} (DSP 3) (Fig 7). The total nitrification rate for DSP 1 were much higher than rates found by Caicedo et al. (2004a) in previou experience (112 – 1190 mg N m^{-2} d^{-1}) with the same experimental system, but fed with a effluent of a sewage fed UASB reactor. A possible explanation for this difference may b the low loading rates in the system due to the high organic removal efficiency of th laboratory scale UASB reactor used in the present experiments. Only Pond 1 presented medium loading rate of 184 Kg BOD_5 $ha^{-1}d^{-1}$. All other ponds presented loading rate between 66 and 32 Kg BOD_5 $ha^{-1}d^{-1}$. The combined analysis of nitrification rates ar organic loading rates for the three experiments demonstrate that when organic loading rate were below 70 Kg BOD_5 $ha^{-1}d^{-1}$, nitrification rates showed a considerable increas reaching values above 1000 mg N m^{-2} d^{-1}. As a consequence of higher total nitrificatio rates and lower denitrification rates, nitrate accumulation was more pronounced in th study than in previous experiences (Caicedo et al., 2004a).

A multiple linear regression with step wise selection performed between nitrification ra vs. oxygen and ammonium nitrogen concentration oxygen and ammonium nitroge concentration eliminated ammonium nitrogen as a variable. Nitrification rates for DSP and 2 show a significant linear relation (p= 0.05) with top layer oxygen concentratio present in the ponds (nitrification rate = 3270 x oxygen concentration; nitrification rate i mg N m^2 d $^{-1}$, oxygen concentration in mg l^{-1}, adjusted R^2 = 0.94). Similar results wer obtained using average oxygen concentration. In DSP 3, nitrification rates observed wer lower than for the other two DSPs, in spite of the higher oxygen level reached in th system. The availability of ammonium nitrogen was lower in this experiment, especially i the last ponds, but concentration were above 1 mg NH_4^+-N l^{-1}, the half saturatio concentration coefficient proposed by Henze et al. (1986, as cited by Water Environment. Federation, 1998). It is therefore more likely that nitrification rates in this experiment wer lower due to unfavorable pH values (pH 6.8 – 4.5). The optimum pH range for nitrificatio reported in literature is 7.5-8.5 (Metcalf and Eddy, 1991). As the nitrification proces consumes alkalinity, the pH may drop if there is not enough buffer capacity. Whe denitrification process is present, it partially restores the level of alkalinity in the syster This shows the importance of having a good pH control when nitrification an denitrification processes are involved. Nevertheless, the relative amount of influe nitrogen that was oxidized in DSP 3 (67%) was higher than in DSP 1 (57%) and DSP (59%).

Net nitrification rates obtained in DSP 1, 2, and 3 were 41–2763 mg $Nm^{-2}d^{-1}$, 17–3032 m N m^{-2} d^{-1} and 46–1485 mg N m^{-2} d^{-1}. These results are much higher than reported values i literature (Eighmy and Bishop, 1989; Zimmo, 2004; Caicedo et al., 2004a). Nitra accumulation could occur because the systems presented good conditions for th nitrification process to pursue, while denitrification process was inhibited by the lack o organic matter.

From the previous discussions, some applications may be proposed. A system combinin anaerobic pre-treatment and a duckweed system with 0.4 m depth with more than seve ponds in series seems to offer the best option for several reasons:

- Shallow ponds are very easy to build, to operate and to maintain and they can be regarded mostly as a crop production system.
- Shallow ponds will have small storage capacity for sediments. The anaerobic pre-treatment will reduce the amount of organic matter and solids entering the pond system.
- If the objective of the treatment is recovery of nitrogen then the stimulation of duckweed incorporation and the reduction of effluent nitrogen to a suitable range for irrigation would be the best option. In this case it would be recommendable a strategy to reduce denitrification. The configuration of an efficient anaerobic pre-treatment followed by a series of ponds will allow nitrification process and minimize denitrification process. Duckweed is known to prefer ammomium as a source of nitrogen, but it can use nitrates when ammonium nitrogen has been exhausted (Porath and Pollock, 1982).
- In ponds with plug flow hydraulic conditions a lack of nutrients may occur towards the last part of the system; by-pass feeding to and intermediate point of the system will be recommendable as shown in Figure 8. This bypass feeding should be small percentage of the influent flow to supply some nutrients without deterioration of treatment efficiencies.
- If the objective of the treatment is nitrogen removal due to disposal or irrigation regulations, recycling of final effluent to the UASB reactor to stimulate denitrification would be an interesting option.

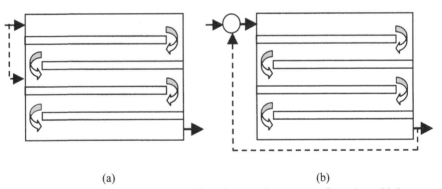

(a)   (b)

Fig 8. Schematic diagrams of (a) proposed plug flow duckweed system configuration with by-pass feeding and (b) proposed plug flow duckweed system with effluent recirculation to a UASB reactor.

Conclusions
- Average total nitrogen removal rates were 598 mg N m^{-2} d^{-1} for DSP 1, 589 mg N m^{-2} d^{-1} for DSP 2 and 482 mg N m^{-2} d^{-1} for DSP 3. In spite of the lower nitrogen removal rate in DSP 3, it has the higher removal efficiency (44 %, 43 % and 62 % for DSP 1, 2 and 3 respectively) due to the lower surface loading rate in this system. This shows that using the percentage of removal as a parameter for comparison should be done with care and the operational parameters of the compared systems should be taken into account. .

- Denitrification was the most important nitrogen removal mechanism for the three systems (56 %, 57 % and 38 % of total influent nitrogen for DSP 1, 2 and 3 respectively), although it was probably limited by BOD_5 supply. Rates from first to last pond were in the range of $820 - 70$ mg N m^{-2} d^{-1} (DSP 1), 547-115 mg N m^{-2} d^{-1} (DSP 2), 681-19 mg N m^{-2} d^{-1}(DSP 3).
- Nitrogen removal via biomass production was the second removal mechanism for the three experiments (15 %, 15 % and 22 % for DSP 1, 2 and 3 respectively). Biomass productions were in the range of 113–65, 106–60, 124–48 g m^{-2} d^{-1} (fresh weight) for DSP 1, 2 and 3 respectively. Nitrogen biomass up-takes were in the ranges of 312-161, 305-155, 302-105 mg N m^{-2} d^{-1} for Exp. 1, 2 and 3 respectively.
- Biomass production was reduced significantly in the last ponds of DSP 3, most probably because of the low pH in those ponds. pH and alkalinity are important parameter to be controlled when nitrification occurs in the duckweed system.
- Pond depth does not seem to determine nitrification or denitrification. Nitrification seems to be related to surface organic loading rate, while denitrification was related to BOD availability.
- The comparison between two pond systems with different depths, but operated at the same hydraulic surface loading rate (DSP 1 and 2) showed similar nitrogen removals in the shallower system as in the deeper system. This suggests that duckweed pond system could be designed with the shallower depth without affecting surface loading and nitrogen removal efficiency. This also shows that nitrogen removal is governed by the surface loading rate rather than by the hydraulic retention time.

Acknowledgements
The authors would like to thank Mr. D. Agudelo and Ms. Z. Palacios for their help in the collection of the data, to Ms. I. Yoshioka for her collaboration in the statistical analysis, to the Netherlands's Government and Universidad del Valle for their support in the development of this research through the SAIL ESEE project.

References
Alaerts G., Mahbubar Rahman, Kelderman P. (1996). Performance Analysis of a full-scale duckweed-covered sewage lagoon. *Wat. Res.* 30 (4), 843-852.

A. P. H. A. (1995). American standard methods for the examination of water and wastewater. 19th edition. New York.

Brock T. D., Madigan M., Martinko J., Parker J. (1991). Biology of Microorganism. 17st Edition. Prentice Hall Inc., London.

Caicedo J. R., Steen N. P. van der, Arce O., Gijzen H J. (2000). Effect of total ammonium nitrogen concentration and pH on growth rates of duckweed (*Spirodela polyrrhiza*). *Wat. Res.* 34(15), 3829-3835.

Caicedo J. R., Steen N. P., Gijzen H. J. (2003). The effect of anaerobic pre-treatment on the performance of duckweed stabilization ponds. Proceedings of International Seminar on natural systems for wastewater treatment. Agua 2003. Cartagena Colombia.

Caicedo J. R., Steen N. P. van der, Gijzen H.J. (2004a). Nitrogen balance of duckweed covered sewage stabilixation ponds. In this thesis.

Caicedo J. R., Steen N. P. van der, Gijzen H.J. (2004b). Effect of introducing aerobic zones into a series of duckweed stabilization ponds on nitrification and denitrification. In this thesis.

Craggs R. J., Davies-Colley R. J., Tanner C. C., Sukias J. P. (2002). Advanced pond system: performance with high rate ponds of different depths and areas. 5th International IWA Speicalist group conference on waste stabilisation ponds. Pond Technology for the new millennium. Volume 1. Auckland, New Zealand. April.

Daniel W. W. (1990). Applied nonparametric statistics. Georgia State University. Houghton Mifflin Company. Boston

Eighmy T. T. and Bishop P. L. (1989). Distribution and role of bacterial nitrifying populations in nitrogen removal in aquatic treatment system. *Wat. Res.* 23 (8), 947-955.

Gijzen H.J. and Ikramullah M. (1999). Pre-feasibility of duckweed-based wastewater treatment and resource recovery in Bangladesh. World Bank Report, Washington D. C.

Kvet J., Rejmankova E., Rejmanek M. (1979). Higher aquatic plants and biological wastewater treatment. The outline of possibilities. Aktiv Jihoceskych vodoh Conf. Pp 9.

Landolt E. and Kandeler R. (l987). The family of *Lemnaceae* monographic study, Vol.2. *Veroeffentlichungen des geobotanisches Institutes der ETH*, Stiftung Rubel, 95, Zurich, 1-638.

Mara D. D., Alabaster G. P., Pearson H. W., Mills S. W. (1992). Waste stabilization ponds. A design manual for Eastern Africa. Lagoon Technology International. Leeds, Inglant.

Metcalf and Eddy. (1991). Waste Engineering. Treatment, disposal and reuse. Tchobanoglous G. and Burton F. L. [eds.]. 2nd Ed. McGraw Hill, Inc. USA.

Oostrom A. J. van (1995). Nitrogen removal in constructed wetlands treating nitrified meat processing effluent. *Wat. Sci. Tech.* 32 (3), 137-147.

Porath D. and Pollock J. (l982). Ammonia stripping by duckweed and its feasibility in circulating aquaculture. *Aquat. Bot.* 13, 125-131.

Reddy K.R. and DeBusk W.F. (1985) Nutrient removal potential of selected aquatic macrophytes. *Jour. Environ. Qual.* 14(4), 459-462.

Silva S. A., de Olivieira R., Soares J., Mara D. D., Pearson H. W. (1995). Nitrogen removal in pond systems with different configurations and geometries. *Wat. Sci. Tech.* 31 (120), 321-330.

Steen N. P. van der, Brenner A., Oron G. (1998). An integrated duckweed algae po system for nitrogen renoval and renovation. *Wat. Sci. Tech.* 38(1), 335-343.

Zimmo O., Steen N. P. van der, Gijzen H. J. (2003a). Nitrogen mass balance over pi scale algae and duckweed-based wastewater stabilization ponds. In: Nitrog transformation and removal mechanism in algal and duckweed waste stabilizacion pon Ph. D. Dissertation. Wageningen University & International Institute of Hydraulic a Environmental Engineering. Holland

Zimmo O. (2003b). Nitrogen transformation and removal mechanism in algal a duckweed waste stabilizacion ponds. Ph. D. Dissertation. Wageningen University International Institute of Hydraulic and Environmental Engineering. Holland.

Zimmo O., Steen N. P. van der, Gijzen H. J. (2004). Quantification of nitrification a denitrification rates in pilot-scale algae and duckweed-based waste stabilization ponds. *E Tech.* 25, 273-282.

Chapter 8

Comparison of performance of full scale
duckweed and algae stabilization ponds.

Chapter 8

Comparison of performance of full scale duckweed and algae stabilization ponds.

Abstract
The aim of this study was to compare the performance of a duckweed pond and an algae pond treating effluent of sewage fed UASB reactor under similar environmental and operational conditions. The real scale experimental system was composed of two continuous flow channels. One operated as an algae pond and the other as a duckweed pond (*Spirodela polyrrhiza and Lemna minor*). The volume of each channel was 225 m^3, an average surface area of 322 m^2, L/W ratio= 13.1 and depth of 0.7. The wastewater flow was 19.7 m^3 d^{-1}, for each system and the theoretical hydraulic retention time was 11.5 days. The ponds were monitored for the following parameter: Organic matter (BOD$_5$), total suspended solids (TSS), ammonium nitrogen (NH$_4^+$-N), total Kjeldahl nitrogen (TKN), nitrite nitrogen (NO$_2$-N), nitrate nitrogen (NO$_3$-N), total phosphorus (TP) and faecal coliform (FC). The duckweed pond developed different environmental conditions in terms of pH, temperature and oxygen, compared to the algae pond. These differences are likely to affect treatment efficiency for organic matter and nitrogen. The duckweed pond was more efficient in removing organic matter and the algae pond was more efficient in nitrogen removal. Denitrification accounted for most of the nitrogen removal in the algae and duckweed ponds. The second most important mechanism for nitrogen removal was ammonia volatilization for the algae pond and plant up-take for the duckweed pond. In the design of duckweed pond systems special attention should be paid to the reactor configuration, flow pattern, in order to obtain good contact between water column and the duckweed cover and to reduce the presence of short circuiting and dead zones.

Key words
Ammonium volatilization, denitrification, duckweed, *Lemnaceae,* nitrification, nitrogen balance, nutrient recovery, stabilization ponds, waste water treatment.

Introduction
Conventional stabilization ponds are widely used world wide as a low cost wastewater technology with the ability to reach effective removal of organic matter and pathogens (Metcalf & Eddy, 1991; Mara *et al.*, 1992). Duckweed stabilization ponds are a relatively new alternative for wastewater treatment, which is also low cost and has the possibility to produce biomass rich in high quality protein (Skillicorn *et al.*, 1993; Gijzen & Ikramulla, 1999). Some of the drawbacks of the conventional stabilization ponds are the high area requirement and the high effluent suspended solids concentrations due to the presence of algae, which may exert oxygen demand in the receiving water bodies or may cause clogging of soil and irrigation sprinklers if used for irrigation (Hancock & Buddhavarapu, 1993; Pearson *et al.*, 1995). Nutrient removal efficiency depends very much on the type of pond

and operational conditions. In the case of duckweed ponds, Reed *et al.* (1995) reported poor organic matter removal while Zimmo *et al.* (2002) found duckweed ponds to be more efficient than algae ponds for removal of organic matter and suspended solids. Zimmo et al (2002) also reported that duckweed ponds were less efficient in removing nitrogen and pathogens. Both types of pond systems require large area, but in the case of duckweed ponds, this is not a disadvantage because the ponds can be regarded as a crop field for animal feed production (Skillicorn *et al.*, 1993; Oron, 1994) rather than an area occupied by a treatment system.

Recent research on duckweed ponds (Al Nozaily, 2001; Zimmo, 2003; Caicedo *et al.*, 2003) shows a good potential for this technology as a sustainable alternative for wastewater treatment. Most of the research so far has been performed at laboratory or pilot scale. In the process of technology-development it is important to test findings at full scale. This includes studies to compare duckweed-based ponds and conventional algae ponds operated under similar conditions of climate, configuration, wastewater composition and loading rate. The aim of this study was to compare the performance of a full scale duckweed pond with a full scale algae pond treating effluent of a UASB reactor.

Materials and methods

Experimental set up
The experiment was carried out in the Wastewater Research Station of Ginebra, a small municipality located in southwest of Colombia with about 8.000 inhabitants in the urban area. The village has a tropical climate with an average temperature of 23 °C. The wastewater is typical domestic wastewater, as the town does not have industrial development. The domestic sewage is collected and transported in a sewer to the treatment station, about 2 Km from the town.

The full scale experimental system was composed of two continuous flow channels in parallel receiving the effluent of a UASB reactor. These channels were a sub-division of an earthen maturation pond which was previously operated for several years. This ensured that the soil was already clogged with settled solids reducing the seepage water losses.

One of the channels was operated as an algae pond and the other as a duckweed pond (Fig. 1). The length of the ponds was 65 m. The cross section of the algae pond was rectangular with 4.95 m width, 0.7 m. depth. The cross section of the duckweed pond was trapezoidal with an average width of 4.95 and depth of 0.7 m. The width at the water surface of the duckweed pond was 5.2 m, allowing more surface for duckweed growth. The volume of both systems was 225 m^3 and the L/W ratio was 13.1.

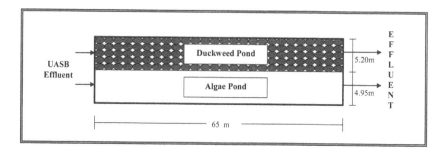

Fig 1. Schematic diagram of the experimental system.

Fig. 2. Overview of the experimental system. Center and left hand side ponds were used in this research.

Operational methods.
The wastewater flow was 19.7 m^3 d^{-1}, for both systems and the theoretical hydraulic retention time was 11.5 days. The duckweed pond was initially seeded with *Spirodela polyrrhiza*. Later on the system developed a mixed culture of *Spirodela polyrrhiza and Lemna minor*. *Lemna minor* was probably brought to the system by birds from nearby water bodies. The surface of the duckweed pond was divided into three equal segments with floating PVC pipes to prevent duckweed from drifting and to maintain a full cover of duckweed on the pond.

The domestic wastewater composition showed hourly changes throughout the day. The fluctuations in composition of the UASB effluent were less pronounced than for the raw wastewater due to the buffering effect in the reactor but they were still significant. To determine the average composition of the UASB reactor effluent, 24-hour sampling programs were conducted. The day was divided in periods of six hours. During each period a composite sample was collected by taking every half an hour a fixed-volume of sample which were pooled and analyzed. Given the long retention time in the pond systems, variations in the effluent of the ponds are not likely and therefore composite samples were collected over a 2 hour period (9-11 am).

The ponds were monitored for the following parameter: Organic matter (BOD_5), total suspended solids (TSS), ammonium nitrogen, total Kjeldahl nitrogen (TKN), nitrite nitrogen (NO_2-N), nitrate nitrogen (NO_3-N), total phosphorus (TP) and faecal coliform (FC). Samples were collected and analyzed every two weeks. The experiment lasted for six months.

Part of the duckweed cover was harvested twice a week. Biomass samples were taken with a strainer of known surface area and were allowed to drain for about 5 minutes. The fresh weight was determined and the density was calculated. Enough biomass was harvested to leave a density of 500 g m^{-2} (fresh weight) after each harvesting. This was enough biomass to generate a closed duckweed cover over the water surface.

Ammonia volatilization.
Ammonia volatilization was determined according to a method adapted from the ones described by Shilton *et al.* (1996) and Zimmo *et al.* (2003a). A Plexiglas transparent aquarium box was placed on the surface water with the open side just below the water surface (Fig. 3). A constant airflow was maintained through this box using a vacuum pump. The air-flow was forced through a column and a conical flask filled with a boric acid solution (2%) where the ammonia was trapped. After 24 hours, the boric acid was titrated with standard 0.02 N H_2SO_4 to calculate the amount of N-NH_3 trapped per day. This value was divided by the surface area (0.237 m^2) of the box to obtain the amount of volatilized ammonia in mg N m^{-2} d^{-1}. The measurements were performed in different locations along each pond to calculate an average volatilization rate (n= 9).

Sediment Sampling and Analysis
The rate of solids accumulation was determined by measuring the depth of the sediment layer at the beginning and at the end of the experiment. To determine the depth of sludge a modification of the method proposed by Van Oostrom (1995) was used which consists of introducing a wooden or plastic stick covered with absorbent cotton lining into the bottom of each pond. Once it reaches the bottom, it should be rotated gently to allow the sediments to stick to the cotton cloth. The stick is taken out carefully and the depth of sludge is measured with a ruler (n=36).

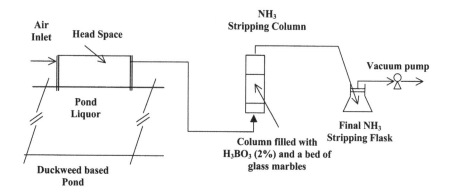

Fig. 3. Diagram of experimental set-up used for measurement of ammonia volatilization.

At the end of the experiment, samples of sludge were taken from the ponds at different places along the each channel with a special sludge pump and their nitrogen content was determined in triplicate. (n=9).

Duckweed Sampling and Analysis
Every month, equal weights of biomass from the different compartments of each pond were mixed all together and 20 grams of fresh weight (FW) were dried in the oven at 70°C for 24 hours to determine the dry weight (n=6). The Kjeldahl nitrogen was determined in the dried biomass in triplicates to establish its nitrogen content (n=6). The nitrogen up-take by the biomass is calculated from the fresh weight biomass production in the system multiplied by its percentage of humidity and by its percentage of nitrogen content.

Nitrogen Balance
The law of conservation of matter was used as the basis for mass balance in this study. The mass balance equation that was used is the following:

$$N_{in} = N_{out} + N_{dw} + N_d + N_v + N_s - N_f \qquad (1)$$

Where,

N_{in} $= (N_{kj} + NO_3^- + NO_2^-)$ in the influent of the duckweed system
N_{out} $= (N_{kj} + NO_3^- + NO_2^-)$ in the effluent of the duckweed system
N_{kj} = Kjeldahl nitrogen (Organic nitrogen + ammonia nitrogen)
N_{dw} = N removed via duckweed harvesting
N_d = N removed via denitrification
N_v = N removed via ammonia volatilisation
N_s = N-accumulation in sediment
N_f = N-fixation

All fluxes are expressed in g d^{-1}. In equation (1), the denitrification flux was assumed as the nitrogen input minus all nitrogen outputs. The nitrification flux was calculated as: denitrification + effluent oxidized nitrogen. Nitrate uptake by duckweed was assumed negligible due to the presence of ammonium nitrogen the preferred source of nitrogen for the duckweed (Porath & Pollock, 1982).

Analytical methods.
Biochemical Oxygen Demand (BOD$_5$), Total Suspended Solids (TSS), Total Kjeldahl Nitrogen (TKN), Ammonium Nitrogen (NH$_4^+$-N), Nitrate Nitrogen (NO$_3$-N) and Nitrite Nitrogen (NO$_2$-N) were measured according to the Standards Methods (APHA, 1995). Unfiltered water samples were analyzed. Dissolved oxygen, pH, and temperature were measured with electrodes. Faecal coliform counts were determined using the membrane filtration method with Cromocult medium from Merck as the growth medium.

Data analysis.
The data were compared and analyzed statistically; results for different treatments lines were compared using ANOVA test and the non-parametric Krusk & Wallis method (Daniel, 1990). The first method assumes independency on the compared data while the second method does not. The results with the two methods were very similar.

Results

Wastewater characteristics.
The average composition of the UASB effluent is presented in Table 1. These are the results of the composite sampling programs run to characterize it.

Table 1. UASB reactor effluent characteristics

Parameter	Units	UASB Effluent *
pH	°C	6.7 ± 0.2
Temperature		24.6 ± 1.7
COD	mg l^{-1}	177 ± 38
BOD$_5$	mg l^{-1}	118 ± 33
TKN	mg l^{-1}	41.8 ± 6.8
N-Ammonium	mg l^{-1}	34.2 ± 7.1
N-Organic	mg l^{-1}	7.6 ± 3.4
N-Nitrites	mg l^{-1}	0.03 ± 0.01
N-Nitrates	mg l^{-1}	<0.01
Total Phosphorus	mg l^{-1}	6.3 ± 1.5
Total Solids	mg l^{-}	363 ± 73
Suspended Solids	mg l^{-1}	50 ± 14
Conductivity	µS cm^{-1}	680 ± 10

* Averages ± S.D. (n=4 daily average; each n = average of 4 integrated samples/day)

Temperature, pH and dissolved oxygen profiles.

Longitudinal and vertical (depths: 7cm, 34 cm, 63 cm) profiles were established during the morning (9-10 am) and afternoon (2-3 pm). Longitudinal profiles did not show great variations of pH, temperature and dissolved oxygen concentrations, along the systems when compared for the same depth. Vertical profiles showed clear differences between duckweed and algae ponds. The vertical gradients are considerably stronger in algae ponds than in duckweed ponds, for all three parameters. The differences are more pronounced during the afternoon (Fig. 4).

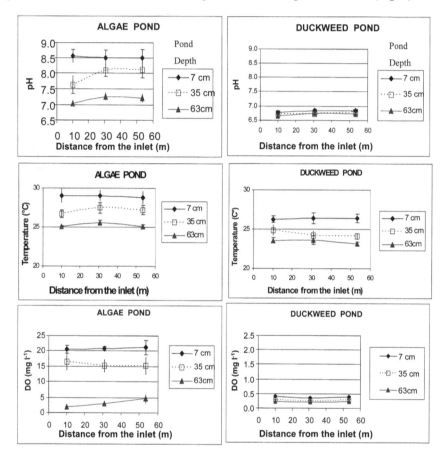

Fig 4. Temperature, pH and dissolved oxygen concentration profiles in the duckweed and algae pond in the afternoon (n= 9). The error bars indicate standard error. The scale for pH and temperature profiles are the same for duckweed and algae pond and different for oxygen profile.

Organic matter removal

In terms of organic matter removal, the UASB reactor removed the greatest percentage of the BOD_5 present in the raw wastewater (62%). The average organic loading rate to the ponds was 72 kg BOD_5 ha^{-1} d^{-1}. The duckweed pond removed

37% of the load received from the UASB reactor and the algae pond removed 11 % (Fig 5). The degradation constants (k_{BOD}) were 0.04 d^{-1} and 0.01 d^{-1} for duckweed pond and algae pond respectively, assuming from a tracer experiment analysis that the ponds behaved as two complete mixed reactors in series. Effluent BOD_5 concentrations were statistically different at 95 % level of confidence. Effluent suspended solids were 52 ± 20 mg l^{-1} in the algae pond and 44 ± 9 mg l^{-1} in the duckweed pond, but not significantly different at the 95 % level.

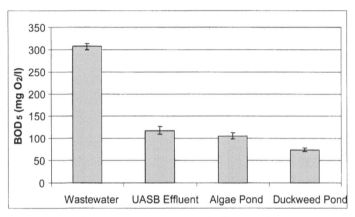

Fig 5. BOD_5 concentration of raw wastewater and effluents of UASB reactor, algae pond and duckweed pond (n= 12). The error bars indicate standard error.

Faecal coliform removal

The geometric mean values of the number of faecal coliforms for the UASB effluent, algae pond and duckweed pond are presented in Figure 6. In the algae pond the FC removal was 1.61 log units and in the duckweed pond the FC removal was 1.39 log units. The faecal coliform die-off coefficient (K_B) was 0.94 d^{-1} for the algae pond and 0.7 d^{-1} for the duckweed pond.

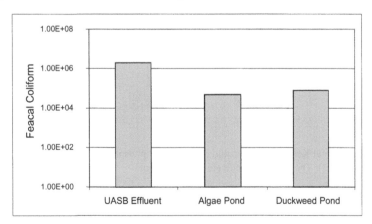

Fig 6.   Faecal coliform removal in the algae and duckweed pond. Values are geometric means (n= 12).

Phosphorus removal

Average phosphorus loading rate was 380 mg P m^{-2} d^{-1} for the two systems. Phosphorus removal was 24% (91 mg P m^{-2} d^{-1}) in the algae pond and 29% (110 mg P m^{-2} d^{-1}) in the duckweed pond. Phosphorus duckweed uptake was 41 mg P m^{-2} d^{-1}. Effluent concentrations in the two systems were found not to be significantly different at the 95% level of confidence.

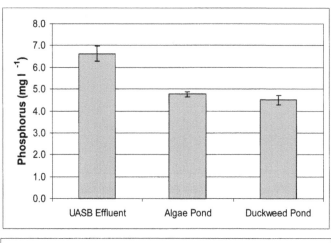

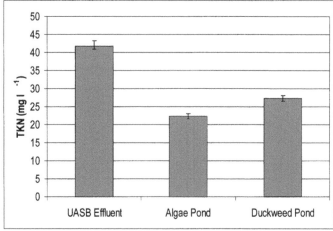

Fig 7. Effluent total Kejdahl nitrogen and effluent total phosphorus in the duckweed and algae ponds (n = 12). The error bars indicate standard error.

Nitrogen removal and balance

Effluent total Kjeldahl nitrogen concentrations in the two systems were significantly different at the 95% level of confidence. Oxidized nitrogen in the effluent was < 0.05 mg l^{-1} in the duckweed pond and < 0.1 mg l^{-1} in the algae pond (n=12).

Nitrogen content of the duckweed was 5.9 % (n=6). Biomass protein content (37%) was calculated by multiplying the nitrogen content by the conversion factor 6.25 (Rusoff *et al.*, 1980). This percentage is in the high side of the range reported in literature (Culley & Epps, 1973; Landolt and Kandeler, 1987; Hammouda *et al.*, 1995; Alaerts *et al.*, 1996)

All nitrogen fluxes considered in equation (1) were calculated for the algae and duckweed ponds as reported in Caicedo *et al.* (2004a) and the results are presented in Figure 8.

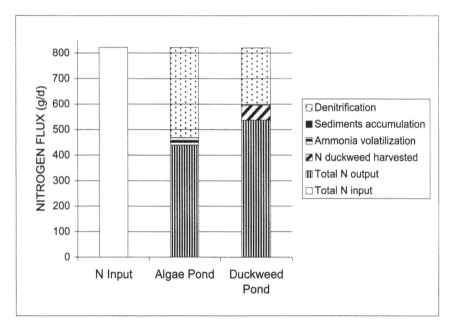

Fig 8. Nitrogen mass balance for the algae and duckweed ponds.

The average nitrogen loading rates to the algae pond and duckweed pond was 2556 mg N m^{-2} d^{-1}. The algae pond achieved significant higher nitrogen removal (46.6%; 1192 mg N m^{-2} d^{-1}) compared with the duckweed pond (34.7%; 967 mg N m^{-2} d^{-1}). The ammonia volatilization was significantly higher in the algae pond (4.1%; 88.4 mg N m^{-2} d^{-1}) than in the duckweed ponds (0.7%; 15 mg N m^{-2} d^{-1}). In the duckweed pond the nitrogen biomass uptake accounted for 7.1% (171.9 mg N m^{-2} d^{-1}) of the removal. In both systems, the removal by sedimentation was < 0.1%. Denitrification accounted for 43.1 % (1101 mg N m^{-2} d^{-1}) of nitrogen removal in the algae pond and for 27 % (688 mg N m^{-2} d^{-1}) in the duckweed pond.

Discussion

Lower levels and less variation of pH, temperature and oxygen were observed in the duckweed pond than in the algae pond. Similar values of these environmental parameters were reported earlier for smaller duckweed ponds in the same research station (Caicedo *et al.*, 2002; Caicedo *et al.*, 2003). Lower values of these parameters were mainly caused by the plant cover which acted as a barrier for light penetration, algal development and oxygen transfer.

Organic matter present in the wastewater was removed mostly in the UASB reactor and as a consequence the influent organic load to the ponds was low compared to permissible organic loads for facultative ponds recommended by Mara (1992). The degradation constants, $k_{BOD} = 0.04$ d $^{-1}$ for the duckweed pond and $k_{BOD} = 0.01$ d $^{-1}$ for the algae pond, were in the lower range of values reported in literature (Fritz, 1985; Saqqar & Pescod, 1996), probably due to the anaerobic pretreatment step in the system. As a result of this pre-treatment, the more recalcitrant fraction of organic matter will remain. The duckweed pond had significantly lower BOD_5 effluent concentration than the algae ponds, which is in agreement with Zimmo *et al.* (2002), who also found better organic matter removal in duckweed ponds than in algae ponds. The removal of total suspended solids was low in both systems. This was reflected in the low accumulation of solids in the bottom of the ponds.

Lower organic matter removal was obtained in the duckweed pond of this study ($k_{BOD} = 0.04$ d $^{-1}$) compared to earlier studies in the same research station at similar retention time, but using a smaller duckweed pilot plant consisting of a series of 3 ponds ($k_{BOD} = 0.13$ d $^{-1}$) (Caicedo *et al.*, 2003). The main reason for this could be the difference in hydraulic characteristics. In the system with the series of ponds the hydraulic pattern tended to be more plug flow, while in the channels used in this study the pattern tended more to mixed flow with the presence of dead zones, which were estimated in a previous tracer study accounted for over 20% of the pond volume (Zimmo *et al.*, 2003b). Assuming this percentage of dead zones, the actual retention time in the system would have been 9.2 days. For the same reasons also lower suspended solids removal could be explained. The horizontal velocity of the water (0.5 cm s^{-1}), probably exceeded the limit of velocity that can drag the fine suspended solids present in the effluent of the UASB reactor which caused re-suspension of the solids (Arboleda, 2002). In the design of duckweed ponds it has been recommended to use high L/W ratios, but this have generated high horizontal water velocities in the systems with the subsequent effect on sedimentation and re-suspension of solids. Improvements in the hydraulic conditions may be obtained by reactor compartmentalization.

Ammonia volatilization rate obtained in this study for the duckweed pond was very similar to rates obtained in previous studies (Caicedo *et al.*, 2004a). This confirms that ammonia volatilization is negligible as nitrogen removal mechanism in duckweed ponds. This is also in agreement with Zimmo *et al.* (2003a) who found rates in the same order of magnitude (< 15 mg N m^{-2} d^{-1}), which represented less than 1.5% of the nitrogen removal. In the algae pond the volatilization rate was around six times higher than in the duckweed pond, which was expected due to the higher pH levels in the pond during the day period. Zimmo *et al.* (2003a) presented considerable lower average rates (< 30 mg N m^{-2} d^{-1}) for algae ponds. The most likely reason could be that their results are annual averages in a region where the four seasons are present.

In the duckweed pond, denitrification rate (688 mg N m^{-2} d^{-1}) was within the same range as previously found in pilot scale plants (Caicedo et al., 2004a). It is interesting to note that the ratio area/volume of the duckweed pond in this study was much lower than in the pilot scale plant. Although less area was available for biofilm development, there was more organic matter supply for denitrification. In the algae pond, denitrification rate (1101 mg N m^{-2} d^{-1}) was in the range of the more aerobic ponds of the pilot plant scale. The concentration of oxidized nitrogen was almost zero in the effluent of both ponds. This means that nitrification and denitrification rates, in each pond, were similar. Comparing the two ponds, nitrification rate and as consequence denitrification rate were higher in the algae pond. As more oxygen was present in the algae pond during the day, more nitrates were produced and were available to be denitrified. This is in agreement with Caicedo et al. (2004c) who found that the introduction of aerobic zones in a duckweed system increased to a great extent the nitrification and denitrification rates. Under similar climatic conditions, Zimmo et al. (2004) also reported higher average denitrification rate in a series of algae ponds (494 mg N m^{-2} d^{-1}) than in a series of duckweed ponds (346 mg N m^{-2} d^{-1}).

The duckweed pond was initially seeded with *Spirodela polyrrhiza* but it developed a mixed culture *Spirodela polyrrhiza – Lemna minor* which reduced the biomass production. As a consequence the nitrogen biomass uptake also decreased when compared to results of a single culture of *Spirodela polyrrhiza* (Caicedo et al., 2004a). N biomass uptake was lower than values reported in literate. Some of the reported results in the literature however are in the same range (Kvet et al., 1979; Reddy & De-Busk, 1985).

Removal of nitrogen in waste stabilization ponds has been reported to vary from negligible to a rather high percentage of incoming nitrogen (Silva et al., 1995). Results depend on the configuration of the system and operational characteristics of the ponds. The percentage of removal of nitrogen in the algae pond in this study (46.6 %).was considerably lower than reported values by Silva, 1982, (81%) and Zimmo et al., 2002 (77%). The main reason for this difference is the higher retention times in the latter studies. This can be confirmed also when comparing nitrogen removal in the duckweed pond (34.7%; 967 mg N m^{-2} d^{-1}) with previous results at pilot scale fed with the same UASB effluent (58%, 805 mg N m^{-2} d^{-1}, Caicedo et al., 2004a). Although removal rate per unit area was higher in the present study, it represented a lower percentage of incoming nitrogen.

In terms of nitrogen concentration the UASB effluent presented a severe degree of restriction for irrigation according with FAO (1985). After the post-treatment in duckweed or algae ponds it may be classified as slight to moderate restriction. Madera et al. (2004) arrived to similar conclusions.

The effluent of both systems did not fulfill the WHO (1989) faecal coliform guidelines for restricted irrigation ($\leq 10^3$ UFC/100 ml), but they were close to the recommended revised guideline, group B1 (Blumenthal et al., 2000). The first order faecal coliform decay constant K_B found in this study, 0.32 d^{-1} for the algae pond and 0.28 d^{-1} for the duckweed pond, in both cases, were in the lower end of the range reported in literature (Saqqar & Pescod, 1992; Pearson et al., 1995; Steen et al., 1999). The low pathogen removal rate may be the result of several factors like low retention time, and poor hydraulic performance (dead zones, short circuiting).

From above discussion and previous results some preliminary recommendations for duckweed ponds design may be defined:

- Compartmentalization of the reactor could improve the pond performance.
- The integrated system, consisting of UASB reactor, duckweed pond and algae pond offers the possibility to remove the different unwanted component in the wastewater and to recover part of the valuable material present in the wastewater in the form of biomass or biogas.
- Further research should be directed toward the definition of the best combination depending on the treatment objectives to meet effluent requirements according with the final disposition either discharge or irrigation.

Conclusions

The duckweed pond developed lower levels of pH, temperature and oxygen than the algae pond with subsequent effects on organic matter, nitrogen and pathogen removals.

The duckweed pond was more efficient to remove organic matter and the algae pond was more efficient to remove nitrogen and pathogens.

In the design of duckweed pond systems special attention should be paid to the reactor configuration, flow pattern, in order to obtain good contact between water column and the duckweed cover and to reduce the presence of short circuiting and dead zones.

It was confirmed at full scale that the first more important nitrogen removal mechanism was denitrification in the algae and duckweed ponds. The second more important mechanism was ammonia volatilization for the algae pond and biomass up-take for the duckweed pond.

Acknowledgements
The authors would like to thanks Ms I. Yoshioka and Ms Luz Adriana Echavarria for their collaboration in the statistical analysis and the Netherlands's Government and Universidad del Valle for their support in the development of this research through the SAIL ESEE project.

References

Alaerts G., Mahbubar Rahman, Kelderman P. (1996). Performance Analysis of a full-scale duckweed-covered sewage lagoon. *Wat. Res.* 30 (4), 843-852.

A. P. H. A. (1995). American standard methods for the examination of water and wastewater. 19th edition. New York.

Arboleda J. (2000). Teoría y práctica de la purificación del agua. 3ª. Edición. Mc Graw Hill. N.Y.

Blumenthal U. J., Mara D. D., Peasey A., Ruiz-Palacios G, Stott R. (2000). Guidelines for the microbiological quality of treated wastewater used in agricultural: recommendations for revising WHO guidelines. Bulletin of the World Health Organization, 78 (9), 1104-1116.

Caicedo J. R., Steen N. P. van der, Arce O., Gijzen H. (2000). Effect of total ammonium nitrogen concentration and pH on growth rates of duckweed (*Spirodela polyrrhiza*). *Wat. Res.* 34(15), 3829-3835.

Caicedo J. R., Espinosa C., Gijzen H. J., Andrade M. (2002). Effect of anaerobic pre-treatment on physicochemical and environmental characteristics of Duckweed based ponds. *Wat. Sci.Tech.* 45(1), 83-89.

Caicedo J. R., Steen N. P. van der, Gijzen H. J. (2003). The effect o anaerobic pre-treatment on the performance of duckweed stabilization ponds. Proceedings of International Seminar on natural systems for wastewater treatment. Agua 2003. Cartagena Colombia.

Caicedo J. R., Steen N. P. van der, Gijzen H.J. (2004a). Nitrogen balance of duckweed covered sewage stabilixation ponds. In this thesis.

Caicedo J. R., Steen N. P. van der, Gijzen H. J. (2004b). Effect of introducing aerobic zones into a series of duckweed stabilization ponds on nitrification and denitrification. In this thesis.

Caicedo J. R., Steen N. P. van der, Gijzen H. J. (2004c). Effect of pond depth on removal of nitrogen in duckweed stabilization ponds. In this thesis.

Daniel W. W. (1990). Applied nonparametric statistics. Georgia State University. Houghton Mifflin Company. Boston.

FAO. (1985). Water quality for agriculture. Irrigation and drainage technical paper No. 29. Rome, Italy.

Gijzen H.J. and Ikramullah M. (1999). Pre-feasibility of duckweed-based wastewater treatment and resource recovery in Bangladesh. World Bank Report, Washington D. C.

Hammouda O., Gaber A., Abdel-Hameed M. S. (1995). Assessment of the effectiveness of treatment of wastewater-contaminated aquatic systems with *Lemna gibba*. *Enzyme and Microbial Tech.* 17, 317-323.

Hancock S. J. and Buddhavarapu L. (1993). Control of algae using duckweed (*Lemna*) systems. In Constructed wetlands for water quality improvement. Ed. Moshiri G. A. CRC Press, 399-406. Boca Raton, Florida.

Kvet J., Rejmankova E., Rejmanek M. (1979). Higher aquatic plants and biological wastewater treatment. The outline of possibilities. Aktiv Jihoceskych Vodoh Conf. Pp 9.

Madera C., Steen N. P. van der, Gijzen H. J. (2004). Comparison of agronomic quality of effluent from conventional and duckweed waste stabilization ponds for reuse in irrigation. Submitted to *Wat. Sci. & Tech.*

Mara D. D., Alabaster G. P., Pearson H. W., Mills S. W. (1992). Waste stabilization ponds. A design manual for Eastern Africa. Lagoon Technology International. Leeds, England.

Metcalf and Eddy. (1991). Waste Engineering. Treatment, disposal and reuse. Tchobanoglous G. and Burton F. L. [eds.]. 2nd Ed. McGraw Hill, Inc. USA.

Oron G. (1994). Duckweed culture for wastewater renovation and biomass production. *Agricultural Water Management*, 26, 27-40.

Oostrom A. J. van (1995). Nitrogen removal in constructed wetlands treating nitrified meat processing effluent. *Wat. Sci. Tech.* 32 (3), 137-147.

Pearson H W., Mara D. D., Arridge H. A. (1995). The influence of pond geometric and configuration on facultative and maturation wate stabilisation pond performance and efficiency. *Wat. Sci Tech.* 31(12), 129-139

Reddy K.R. and De-Busk W.F. (1985). Nutrient removal potential of selected aquatic macrophytes. *Jour. Environ. Qual.*, 14(4), 459-462.

Reed S. C., Middlebrooks E. J., Crites R. W. (1995). Natural systems for waste management and treatment. 2nd Ed. *Mc Graw Hill.* New York.

Rusoff L. L., Blakeney E. W. Jr., Culley D. D. (1980). Duckweed (Lemnaceae Family): A potential source of protein and amino acids. J. Agric. Food Chem. 28, 848-850.

Saqqar M. M. and Pescod M. B. (1992). Modelling coliform reduction in wastewater stabilization ponds. *Wat. Sci. Tech.* 26(7-8), 1667-1677.

Shilton A., Mara D. D., Pearson H. W. (1995). Ammonia volatilisation from a piggery pond. Wat. Sci. Tech., 33 (7), 183-189.

Silva S. A. (1982). On the treatment of domestic sewage in waste stabilization ponds in N.E. Brazil. PhD. Thesis, University of Dundee, UK.

Silva S. A., de Olivieira R., Soares J., Mara D. D., Pearson H. W. (1995). Nitrogen removal in pond systems with different configurations and geometries. *Wat. Sci. Tech.*, 31 (120), 321-330.

Skillicorn P., Spira W., Journey W. (1993). Duckweed aquaculture, a new aquatic farming system for developing countries. *The World Bank.* 76 p. Washington.

Steen N. P. van der, Brenner A., Van Buuren J., Oron G. (1999). Post-treatment of UASB reactor effluent in an integrated duckweed and stabilization pond system. *Wat. Res.* 33(3), 615-620.

Zimmo O., Al-Sa'ed R. M., Steen N. P. van der, Gijzen H. (2002). Process Performance Assessment of algae-based and duckweed-based wastewater treatment systems. *Wat. Sci. Tech.* 45(1), 9i-101.

Zimmo O. (2003). Nitrogen transformation and removal mechanism in algal and duckweed waste stabilization ponds. Ph. D. Dissertation. Wageningen University and IHE. The Netherlands.

Zimmo O., Steen N. P. van der, Gijzen H. J. (2003a). Comparison of ammonia volatilisation rates in algae and duckweed-based waste stabilization ponds   treating domestic wastewater. *Wat. Res.* 37, 45587-4594

Zimmo O., Al-Ahmed M., Steen N. P. van der, Awuah E., Caicedo J., Gijzen H. J. (2003b). Comparison of hydraulic flow pattern in algae and duckweed-based continuous flow pilot-scale waste stabilization ponds. In: Nitrogen transformation and removal mechanism in algal and duckweed waste stabilization ponds. Ph. D. Dissertation. Wageningen University & International Institute of Hydraulic and Environmental Engineering. Holland.

Zimmo O., Steen N. P van der, Gijzen H. J. (2004). Quantification of nitrification and denitrification rates in pilot-scale algae and duckweed-based waste stabilization ponds. *Env. Tech.* 25, 273-282.

Gijzen H.J. and Ikramullah M. (1999). Pre-feasibility of duckweed-based wastewater treatment and resource recovery in Bangladesh. World Bank Report, Washington D. C.

Hammouda O., Gaber A., Abdel-Hameed M. S. (1995). Assessment of the effectiveness of treatment of wastewater-contaminated aquatic systems with *Lemna gibba*. *Enzyme and Microbial Tech.* 17, 317-323.

Hancock S. J. and Buddhavarapu L. (1993). Control of algae using duckweed (*Lemna*) systems. In Constructed wetlands for water quality improvement. Ed. Moshiri G. A. CRC Press, 399-406. Boca Raton, Florida.

Kvet J., Rejmankova E., Rejmanek M. (1979). Higher aquatic plants and biological wastewater treatment. The outline of possibilities. Aktiv Jihoceskych Vodoh Conf. Pp 9.

Madera C., Steen N. P. van der, Gijzen H. J. (2004). Comparison of agronomic quality of effluent from conventional and duckweed waste stabilization ponds for reuse in irrigation. Submitted to *Wat. Sci. & Tech.*

Mara D. D., Alabaster G. P., Pearson H. W., Mills S. W. (1992). Waste stabilization ponds. A design manual for Eastern Africa. Lagoon Technology International. Leeds, England.

Metcalf and Eddy. (1991). Waste Engineering. Treatment, disposal and reuse. Tchobanoglous G. and Burton F. L. [eds.]. 2nd Ed. McGraw Hill, Inc. USA.

Oron G. (1994). Duckweed culture for wastewater renovation and biomass production. *Agricultural Water Management,* 26, 27-40.

Oostrom A. J. van (1995). Nitrogen removal in constructed wetlands treating nitrified meat processing effluent. *Wat. Sci. Tech.* 32 (3), 137-147.

Pearson H W., Mara D. D., Arridge H. A. (1995). The influence of pond geometric and configuration on facultative and maturation wate stabilisation pond performance and efficiency. *Wat. Sci Tech.* 31(12), 129-139

Reddy K.R. and De-Busk W.F. (1985). Nutrient removal potential of selected aquatic macrophytes. *Jour. Environ. Qual.,* 14(4), 459-462.

Reed S. C., Middlebrooks E. J., Crites R. W. (1995). Natural systems for waste management and treatment. 2nd Ed. *Mc Graw Hill.* New York.

Rusoff L. L., Blakeney E. W. Jr., Culley D. D. (1980). Duckweed (Lemnaceae Family): A potential source of protein and amino acids. J. Agric. Food Chem. 28, 848-850.

Chapter 9

Summary

Chapter 9

Summary

There is a diversity of conventional technologies available for removal of pollutants from wastewater. Most of these technologies are aerobic alternatives with high construction cost and high energy consumption and require skilled personal for operation and maintenance. As a consequence, only countries with a high gross national product (GNP) can afford these options. Where these technologies were introduced in developing countries, in most cases these could not be operated sustainably, leading to loss of investments and continued water resource contamination. Extensive investments in wastewater treatment plants world-wide during the last decades have greatly reduced the organic loading of receiving water bodies in high GNP countries. Only recently, many of these plants were appropriated to remove nitrogen and phosphorus. The increasing use of chemical fertilizer may cause high levels of eutrophication in water bodies, which may induce algae blooms resulting in strong fluctuations in oxygen concentration. Oxygen depletion causes fish kill as well as odor problems.

The situation in countries with a low GNP is worse than in the developed world. The unequal expansion of water supply coverage compared to the expansion in wastewater and sanitation services leads to increased contamination of surface and ground waters. The general trend is to use conventional WWT systems for big cities, but for medium and small sized cities non-conventional systems are often considered. Therefore, there is an urgent need to develop and improve low cost technologies for wastewater treatment that are within the economic and technological capabilities of developing countries. In countries like Colombia it is very common that the regulation controls mainly the removals of organic matter and suspended solids. Other parameters like nitrogen, phosphorus, pathogens, micro-contaminants are also crucial and need to be addressed. This makes a response via conventional technologies very expensive, and for developing regions in fact unachievable. It would be ideal if new technologies can provide besides the removal of organic matter and solids, resource recovery like the generation of biogas (energy production) or high quality biomass (animal fodder). At the moment, no technological packages appear to be readily available.

Experience has shown that no single technology can offer an optimum treatment for the different components to be treated in wastewater or to recover them as valuable resources. Therefore an adequate combination of different technologies in an integrated system could convert a wastewater treatment into an attractive sustainable system. For example UASB reactor and duckweed ponds are relatively low cost technologies and their combination offers several advantages. Firstly, anaerobic treatment will reduce considerably the organic matter in the wastewater and convert it into methane, which can be used as a source of renewable energy. Secondly, the effluents of anaerobic treatment could be post-treated to meet discharge standards in duckweed ponds for nutrient recovery in the form of high quality biomass. At this point three valuable products can be listed: biogas for use as an energy source, biomass that can be used for aquaculture or animal feed and treated effluent that can be re-used in irrigation. A system that generates such by-products increases the feasibility and sustainability of pollution control programs. Furthermore, the products may help to address the increasing need for food production in the world.

The development of duckweed pond technology has been concentrated on the study of the processes occurring within the ponds, with respect to organic matter, nitrogen, phosphorus and pathogen removal and the corresponding mechanisms. Further research is needed in order to have a good control of effluent nitrogen levels. There are still important questions to be answer like how to maximize nitrogen recovery via duckweed production, how to get good effluent levels depending on effluent reuse. If the effluent is going to be used in crop irrigation, to reduce nitrogen effluent concentration to 15-20 mg l^{-1} will be enough. If the effluent is going to be discharge in surface waters the nitrogen level would have to be reduced as much as possible. Therefore it is important to study how the design and combination of technologies could generate the required nitrogen effluent levels. The present work was focus on the study of the effect of different operational variables, like the effect of anaerobic pre-treatment, the combination of algae and duckweed ponds, the effect of pond depth on nitrogen transformation and removals.

The effect of anaerobic pre-treatment on environmental and physicochemical characteristics of duckweed stabilization ponds was studied in **Chapter 2**. The environmental and physicochemical conditions affect both plant growth and microbiological treatment processes in the system. Two series of continuous-flow pilot plants, composed of seven ponds in series each, were operated side by side. One system received artificial sewage with anaerobic pre-treatment, while the other system received the same wastewater without anaerobic pretreatment. pH, temperature, dissolved oxygen, alkalinity, conductivity, biochemical oxygen demand, total and ammonium nitrogen, nitrites and nitrates, and phosphorus were monitored under steady state conditions. It was found that pH levels were very stable in both systems with and without anaerobic pretreatment. Vertical temperature gradients were present during daytime but not as strong as they may occur in conventional stabilization ponds. Oxygen levels were significantly higher in the duckweed system with anaerobic pretreatment, especially in the top layer. (up to 2 mg O_2 l^{-1}) than in the system without pretreatment (up to 1.2 mg O_2 l^{-1}). Nevertheless, aeration rates were low in both systems. Both systems were efficient in removing organic matter. The system without pretreatment obtained 98% of BOD_5 removal in pond 4, so 12 days of retention time will be sufficient to reach high organic matter removal. The system with pretreatment obtained also 98% BOD_5 removal (92% in UASB reactor). In this case the duckweed ponds will serve as a polishing step for remaining organic matter. Nutrient removals were 37-48% for nitrogen and 45-50 % for phosphorus in the lines with and without pretreatment respectively.

The main form of nitrogen in anaerobic effluent is ammonium. This is the preferred nitrogen source for duckweed, but at high levels it may become inhibitory to the plant. Renewal fed batch experiments at laboratory scale were performed (**Chapter 3**) to assess the effect of total ammonia (NH_3 + NH_4^+) nitrogen and pH on the growth rate of the duckweed *Spirodela polyrrhiza*. The experiments were performed at different total ammonia nitrogen concentrations, different pH ranges and in three different growth media. The inhibition of duckweed growth by ammonium was found to be due to a combined effect of ammonium ions (NH_4^+) and ammonia (NH_3), the relative importance of each one depending on pH.

The effect of anaerobic pre-treatment on the performance of a duckweed stabilization pond system was assessed in a pilot plant located in the Ginebra Research Station-Colombia (**Chapter 4**). The pilot plant consisted of two lines of seven duckweed ponds in series. One line received de-gritted domestic wastewater and the other received effluent of a 250 m^3 Up-flow Anaerobic Sludge Blanket (UASB) reactor, treating the same wastewater. Both lines were operated at a total hydraulic retention time of 21 days. The systems were monitored for temperature, pH, oxygen, biochemical oxygen demand, chemical oxygen demand, total suspended solids, total phosphorus, biomass production, and different forms of nitrogen. No effect of anaerobic pretreatment was observed on pH and temperature in the two systems. Oxygen concentrations were higher in the system with UASB reactor. Although both systems complied with the Colombian regulation for BOD removal (> 85%), pretreatment with UASB reactor may contribute to the reduction of area requirement for the stabilization ponds. Effluent quality in terms of total suspended solids was excellent, i.e. 9 ± 2 and 4 ± 1 mg l^{-1} in the system with and without pre-treatment, respectively. Total nitrogen removals were 63 % and 68% and phosphorus removals were 24% and 29% in the system with and without pre-treatment, respectively. The differences between the two systems were found not to be significant. Duckweed biomass production was in the range of 54-90 g m^{-2}-d^{-1} (fresh weight) in the system with pre-treatment and 36-84 g m^{-2}-d^{-1} in the system without pre-treatment. Total biomass productions were significantly different at 92% level of confidence. Protein content was 35.1% and 36.6% for the system with and without pre-treatment, respectively.

Nitrogen removal is nowadays one of the most important effluent treatment objectives because of the serious pollution problems it causes to the environment. How nitrogen is transformed and removed in duckweed ponds was studied and nitrogen balances were established (**Chapter 5**). The experimental system was the same as in the previous chapter. Ammonia volatilization was found to be not an important removal mechanism in duckweed ponds (less than 1%). Removal by sedimentation was also low at 2.1% and 4.7% for the systems with and without anaerobic pre-treatment, respectively. Instead, denitrification was found to be the most important removal mechanism (42% and 48%), followed by duckweed biomass up-take (15.6% and 15.1%). Average nitrogen biomass up-take rates were 199 mg N m^{-2} d^{-1} and 193 mg N m^{-2} d^{-1} for the system with and without pre-treatment, respectively. Nitrification rates were in the range of $112-1190\,mg\,N\,m^{-2}\,d^{-1}$

and 58-1123 mg N m^{-2} d^{-1} for the system with and without anaerobic pretreatment respectively. Denitrification rates were in the range of 112 – 937 mg N m^{-2} d^{-1} and 59 – 1039 mg N m^{-2} d^{-1} for the system with and without pre-treatment respectively. The configuration of the system, in particular the down and up flow pattern seemed to have an important stimulating effect on denitrification rates, probably by causing alternative exposure of the pond water to aerobic and anoxic conditions.

Although the potential of duckweed ponds for removing carbonaceous and suspended material from wastewater has been demonstrated, the system could be further optimized for nitrogen removal. The effect of introducing algae-ponds (aerobic zones) into a series of duckweed stabilization ponds on nitrification and denitrification (**Chapter 6**) was studied in two consecutive phases. During the first phase, the seven ponds of the pilot plant were fully covered with duckweed (*Spirodela polyrrhiza*). Before the start of the second phase, the duckweed cover was removed from ponds 1 and 3, with a view to allow algae growth in the 'open' ponds. The feed of the duckweed pond system consisted of the effluent of a real scale UASB reactor, which treated domestic wastewater from Ginebra-Colombia. The system was operated with a continuous flow to produce a hydraulic retention time (HRT) of 3 days per pond and a total HRT of 21 days. Effluent total nitrogen was significantly different in the two phases, with 13.8± 2.9 mg TN l^{-1} (63 % removal) and 3.7±1.5 mg TN l^{-1} (90%) for first and second phase, respectively. Denitrification was the most important removal mechanism during both phases, and amounted to 43.5 % and 76.2 % of influent nitrogen, in first and second phase, respectively. Ammonia volatilization and sedimentation were insignificant processes for nitrogen removal in both phases. Nitrification played an important role in nitrogen transformations in the duckweed systems and it was favored by the introduction of aerobic zones in ponds 1 and 3. Denitrification also played a key role in nitrogen transformations and removal. Despite the presence of oxygen in the water column, denitrification occurred, probably due to the anaerobic microenvironment of system biofilms. Higher nitrogen removal might be obtained in duckweed pond systems through the introduction of aerobic zones in early stages of the system. Where effluents cannot be reused for crop irrigation, strict nitrogen effluent criteria can be met using hybrid duckweed-algal ponds at considerably shorter hydraulic retention time compared to fully duckweed covered systems.

The effect of pond depth on nitrogen removal in duckweed stabilization ponds was studied in **Chapter 7**. The pilot plant consisted of two lines with seven duckweed ponds in series, with different depths and fed with effluent of a laboratory scale UASB reactor. Three experimental conditions were studied: DSP1 with pond depth 0.7 m and HRT= 21 days, DSP2 with pond depth 0.4 m and HRT = 12 days, and DSP3 with pond depth 0.4 m and HRT = 21 days. The systems were monitored for pH, temperature and oxygen profiles, organic matter removal (BOD$_5$), nitrogen transformations, biomass production and biomass nitrogen content. Average total nitrogen removal rates were 598 mg N m^{-2} d^{-1} for DSP 1, 589 mg N m^{-2} d^{-1} for DSP 2 and 482 mg N m^{-2} d^{-1} for DSP 3. In spite of the lower nitrogen removal rate in DSP 3, it had higher removal efficiency (44 %, 43 % and 62 % for DSP 1, 2 and 3 respectively) due to the lower surface loading rate in this system. This shows that using the percentage of removal as a parameter for comparison should be done with care and the operational parameters of the compared systems should be taken into account. Denitrification was the most important nitrogen removal mechanism for the three DSPs. Nitrogen removal via biomass production was the second most important removal mechanism for the three experiments. Pond depth does not seem to determine nitrification or denitrification. Nitrification seems to be related to surface organic loading rate, while denitrification was related to BOD availability. The comparison between two pond systems with different depth, but operated at the same hydraulic surface loading rate (DSP 1 and 2) showed similar nitrogen removals in the shallower system as in the deeper system. This suggests that duckweed pond system could be designed with shallow depth without affecting surface loading and nitrogen removal efficiency. Nitrogen removal appeared to be governed by surface loading rate rather than by hydraulic retention time.

Most of the research so far has been performed at laboratory or pilot scale. In the process of technology-development it is important to test findings at full scale. In **Chapter 8**, the performance of a full scale duckweed pond was compared with a full scale algae pond treating effluent of a UASB reactor operated under similar conditions of climate, configuration, wastewater composition and loading rate. The real scale experimental system was composed of two continuous flow channels. One operated as an algae pond and the other as a duckweed pond (*Spirodela polyrrhiza and Lemna minor.*). The volume of each channel was 225 m^3, an average surface area of 322 m^2, L/W ratio= 13.1 and depth of 0.7 m. The wastewater flow was 19.7 m^3 d^{-1}, for each system and the theoretical hydraulic retention time was 11.5 days. The ponds were monitored for the following parameters: Organic matter (BOD_5), total suspended solids (TSS), ammonium nitrogen (NH_4^+-N), total Kjeldahl nitrogen (TKN), nitrite nitrogen (NO_2-N), nitrate nitrogen (NO_3-N), total phosphorus (TP) and faecal coliform (FC). The duckweed pond developed different environmental conditions in terms of pH, temperature and oxygen, compared to the algae pond. The duckweed pond was more efficient in removing organic matter and the algae pond was more efficient in nitrogen removal. Denitrification accounted for most of the nitrogen removal in the algae and duckweed ponds. The second most important mechanism for nitrogen removal was ammonia volatilization for the algae pond and plant up-take for the duckweed pond. In the design of duckweed pond systems special attention should be paid to the reactor configuration and flow pattern in order to obtain good contact between water column and the duckweed cover and to reduce hydraulic problems.

Practical applications.
Wastewater treatment can be converted into an attractive, feasible and sustainable alternative by combining anaerobic pretreatment, duckweed ponds, and algae ponds. The integrated system UASB reactor, algae pond and duckweed pond offers the possibility to remove the various unwanted component in wastewater and to recover part of the valuable material present in the wastewater in the form of biomass or

biogas The effluents may be suitable for discharge or for irrigation depending on the removal efficiencies of the system.

The design and operation of this integrated system may have two different approaches. Firstly, one could optimize nitrogen recovery by duckweed uptake and effluent irrigation. Secondly, one could maximize nitrogen removal in order to protect the receiving water resources.

If the objective of the treatment is recovery of nitrogen then the stimulation of duckweed incorporation and the reduction of effluent nitrogen to a suitable range for irrigation would be the best option. The configuration of an efficient anaerobic pre-treatment followed by a series of ponds completely covered with duckweed would be recommendable. Influent ammonium nitrogen concentration below 50 mg l^{-1} and pH below 8 would be desirable to avoid biomass growth inhibition. The comparison between two pond systems with different depths and the same hydraulic surface loading rate showed similar nitrogen removals in the shallower system as in the deeper system. This means that duckweed pond system could be designed with the shallower depth without affecting nitrogen removal efficiency. Shallow ponds are easier to build, to operate and to maintain and in the case of duckweed covered ponds, they can be regarded as a crop production system.

If the objective of the treatment is nitrogen removal due to disposal regulations, a strategy to enhance denitrification should be adopted. Higher nitrogen removals may be obtained in duckweed pond systems through the introduction of aerobic zones in early stages of the system, which allows a considerable reduction of the hydraulic retention time. Strict nitrogen effluent criteria can therefore be met at relatively short hydraulic retention times. The configuration of the system, in particular the down and up flow pattern seems to have an important positive effect on denitrification rates.

Compartmentalization of the treatment system improves the pond performance. In the design of pond systems special attention should be paid to the reactor configuration and hydraulic flow pattern, good contact water-biomass and to avoidance of short circuiting and dead zones.

In the process of technology development the following studies are envisaged and recommended for further research:
- Future studies should be focused on shallow ponds with the views to enhance nitrogen removal via its recovery in the form of duckweed biomass. Shallow ponds will also reduce construction cost of the treatment systems.
- Alternative uses of treated effluent and produced biomass should be investigated. In the case of effluent reuse on irrigation, the reduction of nitrogen concentrations in the treatment system to 15-25 mg l^{-1} will be enough. The use of vegetable biomass as a food complement on the diet of

fish and pork is an alternative that has been preliminary explored in the area of research. Further studies are necessary to determine its feasibility.

- For safe discharge of effluent to open water bodies, effluent nitrogen concentration should be low. In this case nitrogen removal processes may be influence by affecting growth conditions of nitrifiers/dentrifiers like oxygen levels or availability of area for bacterial attachment. It is important to performed studies in order to find the best combination of duckweed and algae ponds for nitrogen removal. The introduction of baffles on the treatment channels will increase the availability of area for biomass growth and will improved the hydraulic characteristics of the treatment systems. The appropriated number and distribution of baffles should be investigated. Recycling of final aerobic effluent to the UASB reactor or to the entrance of the duckweed pond could be an interesting option to stimulate denitrification.

- Pathogen removal will be affected by the use of low pond depths, the presence of aerobic zones and compartmentalization in the treatment system. These effects should be researched in order to optimize also the removal of pathogenic microorganisms.

Chapter 10

Samenvatting

Chapter 10

Samenvatting

Er is een verscheidenheid aan conventionele technologieën voor de verwijdering van vervuilende stoffen in afvalwater. De meeste van deze technologieën zijn aërobe alternatieven met hoge constructie kosten, een hoog energieverbruik en ze vereisen geschoold personeel voor de bedrijfsvoering en het onderhoud. Dit heeft tot gevolg dat alleen landen met een hoog nationaal product (GNP) zich deze opties kunnen veroorloven. Waar deze technologieën in ontwikkelingslanden werden geïntroduceerd, konden deze in de meeste gevallen niet duurzaam bedreven worden. Dit leidde tot een verlies aan investeringen en een voortdurende vervuiling van waterlichamen. Hoge investeringen in afvalwaterzuiveringen gedurende de afgelopen decaden hebben wereldwijd geleid tot een sterke afname in de organische belasting van ontvangende waterlichamen in landen met een hoog GNP. Pas recent zijn de meeste van deze installaties aangepast voor verwijdering van stikstof en fosfor. Het toenemende gebruik van kunstmest kan leiden tot sterke eutroficatie in oppervlaktewater, wat kan leiden tot sterke algengroei met als resultaat sterke fluctuaties in de zuurstofconcentratie. Een tekort aan zuurstof leidt tot vissterfte en stankontwikkeling.

De situatie in landen met een laag GNP is slechter dan in de geïndustrialiseerde wereld. De ongelijke uitbreiding van watervoorziening in vergelijking tot de uitbreiding van afvalwaterzuivering en sanitatie leidt tot een toename van de vervuiling van oppervlaktewater en grondwater. De algemene trend is om conventionele afvalwaterzuiveringsinstallaties te gebruiken voor grote steden, maar voor kleine en middelgrote steden wordt de toepassing van niet-conventionele systemen vaak overwogen. Het is daarom van groot belang om goedkope technologieën voor afvalwaterzuivering te ontwikkelen en te verbeteren. Deze systemen passen bij de economische en technologische mogelijkheden van ontwikkelingslanden. In landen zoals Colombia is het gebruikelijk dat de wetgeving voornamelijk de verwijdering van organische en zwevende stof regelt. Andere parameters zoals stikstof, fosfor, ziektekiemen en microverontreinigingen zijn ook belangrijk en moeten aangepakt worden. Dit maakt een aanpak via conventionele technologieën erg duur, en voor ontwikkelingslanden in feite onbereikbaar. Het zou ideaal zijn als nieuwe technologieën behalve de verwijdering van organische en zwevende stof ook het terugwinnen van grondstoffen mogelijk zouden maken, zoals door de productie van biogas (energie productie) of hoge kwaliteit biomassa (diervoeder). Op dit moment lijken zulke technologieën niet beschikbaar te zijn.

Ervaring laat zien dat geen enkele losstaande technologie een optimale zuivering kan behalen voor de verschillende componenten in afvalwater of die componenten om te zetten in waardevolle stoffen. Een juiste combinatie van verschillende technologieën in een geïntegreerd systeem kan een afvalwater zuivering veranderen in een aantrekkelijk duurzaam systeem. UASB reactoren en eendekroos vijvers zijn bijvoorbeeld relatief goedkope technologieën en het combineren daarvan levert verschillende voordelen op. Ten eerste, anaërobe zuivering zal de organische stof in het afvalwater aanzienlijk verminderen en omzetten in methaan, wat gebruikt kan worden als een duurzame energiebron. Ten tweede, het anaërobe effluent kan nabehandeld worden om aan lozingseisen te voldoen in eendekroos vijvers om de nutriënten terug te winnen in de vorm van biomassa met een hoge kwaliteit. Drie waardevolle producten kunnen genoemd worden: biogas voor gebruik als energiebron, biomassa dat gebruikt kan worden als vee- of visvoeder en effluent dat hergebruikt kan worden voor irrigatie. Een systeem dat zulke bijproducten geneert zorgt voor een toename van de haalbaarheid en duurzaamheid van programma's voor afvalwaterzuivering. Bovendien, deze bijproducten zouden kunnen helpen bij het voorzien in de toenemende vraag naar voedsel in de wereld.

De ontwikkeling van eendekroos vijvers is vooral gericht geweest op het bestuderen van de processen die plaatsvinden in de vijvers, met betrekking tot organische stof, stikstof, fosfor en pathogenen verwijdering en de verantwoordelijke mechanismen. Verder onderzoek is nodig om de concentratie stikstof in het effluent te beïnvloeden. Er zijn nog belangrijke onbeantwoorde vragen zoals hoe de terugwinning van stikstof door eendekroos productie gemaximaliseerd kan worden en hoe een goede kwaliteit effluent voor hergebruikdoeleinden verkregen kan worden. Als het effluent gebruikt zal gaan worden voor irrigatie van gewassen, zal het voldoende zijn om de concentratie stikstof in het effluent te reduceren tot 15-20 mg l^{-1}. Als het effluent geloosd zal gaan worden in oppervlaktewater zal de concentratie stikstof zo veel als mogelijk is verminderd moeten worden. Daarom is het belangrijk om te bestuderen hoe het ontwerp en de combinatie van technologieën de vereiste effluent stikstof concentraties kunnen bewerkstelligen. Dit onderzoek was gericht op het bestuderen van de effecten van verschillende operationele variabelen, zoals het effect van anaërobe voorbehandeling, het combineren van algen en eendekroos vijvers en het effect van de diepte van de vijvers op stikstof omzettingen en verwijdering.

Het effect van anaërobe voorbehandeling op omgevingsfactoren en de fysisch-chemische omstandigheden in eendekroos vijvers is bestudeerd in **Hoofdstuk 2**. De omgevingsfactoren en de fysisch-chemische omstandigheden hebben beiden een effect op de groei van planten en op de microbiologische behandelingsprocessen in het systeem. Twee series van continu bedreven proefinstallaties, elk bestaand uit 7 vijvers in serie, werden naast elkaar bedreven. Eén systeem werd gevoed met anaëroob voorbehandeld kunstmatig rioolwater, terwijl het andere systeem gevoed werd met hetzelfde afvalwater, maar niet voorbehandeld. De pH, temperatuur, zuurstof concentratie, alkaliniteit, geleidbaarheid, biochemisch zuurstof verbruik, totaal en ammonium stikstof, nitriet en nitraat, en fosfor werden gemeten onder steady-state condities. Er werd gevonden dat de pH waarden erg stabiel waren in beide systemen, zowel het systeem met als het systeem zonder anaërobe voorbehandeling. Verticale temperatuur gradiënten waren aanwezig gedurende de dag, maar niet zo sterk als ze wel in conventionele stabilisatie vijvers voor kunnen komen. De concentraties zuurstof waren significant hoger in het eendekroos systeem met anaërobe voorzuivering, in het bijzonder in de bovenste laag (tot 2 mg O_2 l^{-1}) in vergelijking met het systeem zonder voorzuivering (tot 1.2 mg O_2 l^{-1}). Toch was de zuurstof inbreng laag in beide systemen. In beide systemen werd organische stof efficiënt verwijderd. In het systeem zonder voorzuivering werd 98% BOD verwijdering gehaald in vijver nummer 4, dus 12 dagen verblijftijd zal voldoende zijn om een hoge verwijdering van organische stof te bereiken. Het systeem met voorzuivering haalde ook 98% BOD verwijdering (92% in de UASB reactor). In zo'n geval zullen de eendekroos vijvers dienen als een nazuivering voor de verwijdering van organische stof. De verwijdering van nutriënten was 37-48% voor stikstof en 45-50% voor fosfor in de series met en zonder voorzuivering, respectievelijk.

De belangrijkste vorm van stikstof in anaëroob effluent is ammonium. Dit is de stikstof vorm waaraan eendekroos de voorkeur geeft, maar hoge concentraties kunnen remmend werken op de groei. Fed batch experimenten op laboratorium schaal werden uitgevoerd (**Hoofdstuk 3**) om het effect van de concentratie totaal ammonium (NH_3 + NH_4^+) en pH op de groeisnelheid van het eendekroos *Spirodela polyrrhiza* te bepalen. De experimenten werden uitgevoerd met verschillende totaal ammonium concentraties, verschillende pH waarden en in drie verschillende groei media. De remming van de groei van eendekroos door ammonium bleek een gecombineerd effect te zijn van het ammonium ion (NH_4^+) en ammoniak (NH_3), waarbij het relatieve belang van elk afhangt van de pH.

Het effect van anaërobe voorbehandeling op de effectiviteit van een eendekroos vijver is bepaald in een proefinstallatie in het Ginebra Research Station in Colombia (**Hoofdstuk 4**). De proefinstallatie bestond uit twee series van 7 eendekroos vijvers in serie. Eén serie werd gevoed met huishoudelijk afvalwater en de andere met effluent van een 250 m^3 Up-flow Anaerobic Sludge Blanket (UASB) reactor, die hetzelfde afvalwater behandelde. Beide series werden bedreven met een totale hydraulische verblijftijd van 21 dagen. De systemen werden bemonsterd en bemeten voor temperatuur, pH, zuurstof, biochemisch zuurstofverbruik, chemisch zuurstofverbruik, totaal zwevende stof, totaal fosfor, biomassa productie, en de verschillende vormen van stikstof. Er was geen effect van de anaërobe voorzuivering op de pH en de temperatuur in beide systemen. De concentratie zuurstof was hoger in het systeem met UASB reactor. Hoewel beide systemen voldeden aan de Colombiaanse wetgeving voor BOD verwijdering (>85%), kan voorbehandeling met een UASB reactor nuttig zijn voor het reduceren van de benodigde oppervlakte voor de stabilisatie vijvers. De kwaliteit van het effluent met betrekking tot het gehalte totaal zwevende stof was erg goed, namelijk 9 ± 2 en 4 ± 1 mg l^{-1} in het systeem met en zonder voorbehandeling, respectievelijk. De verwijdering voor totaal stikstof was achtereenvolgens 63% en 68% en voor fosfor 24% en 29% in het systeem met en zonder voorbehandeling. Het verschil tussen de twee systemen was niet significant. De productie van eendekroos biomassa varieerde van 54-90 g m^{-2}-d^{-1} (vers gewicht) in het systeem met voorzuivering tot 36-84 g m^{-2}-d^{-1} in het systeem zonder voorzuivering. De totale productie aan biomassa waren significant verschillend bij een 92% betrouwbaarheidsinterval. Het gehalte aan eiwit was 35.1% en 36.6% voor het systeem met en zonder voorzuivering, respectievelijk.

Verwijdering van stikstof is tegenwoordig één van de meest belangrijke doelen voor zuivering, vanwege de ernstige vervuiling van het milieu die het veroorzaakt. Hoe stikstof omgezet en verwijderd wordt in eendekroos vijvers werd bestudeerd en stikstof balansen werden opgesteld (**Hoofdstuk 5**). Het experimentele systeem was hetzelfde als beschreven in het vorige hoofdstuk. Vervluchtiging van ammoniak bleek geen belangrijk verwijderingsmechanisme te zijn in eendekroos vijvers (minder dan 1%). Verwijdering door bezinking was ook gering met 2.1% en 4.7% voor de systemen met een zonder anaërobe voorzuivering, respectievelijk. In plaats daarvan bleek denitrificatie het belangrijkste verwijderingsmechanisme te zijn (42% en 48%), gevolgd door opname door eendekroos (15.6% and 15.1%). De gemiddelde opname van stikstof door biomassa was 199 mg N m^{-2} d^{-1} en 193 mg N m^{-2} d^{-1} Nitrificatie snelheden varieerden tussen 112 – 1190 mg N m^{-2} d^{-1} en 58-1123 mg N m^{-2} d^{-1} voor het systeem met en zonder voorzuivering, respectievelijk. Denitrificatie snelheden varieerden tussen 112 – 937 mg N m^{-2} d^{-1} en 59 – 1039 mg N m^{-2} d^{-1} voor het systeem met en zonder voorzuivering, respectievelijk. De configuratie van het systeem, met name de neerwaartse en opwaartse stroming leek een belangrijk stimulerend effect op de denitrificatiesnelheden te hebben, waarschijnlijk door het veroorzaken van afwisselende blootstelling van het water aan aërobe en anoxische omstandigheden.

Hoewel de geschiktheid van eendekroos vijvers voor het verwijdering van organische en zwevende stof uit afvalwater aangetoond is, kan het systeem nog verder geoptimaliseerd worden voor stikstof verwijdering. Het effect van het introduceren van algen vijvers (aërobe zones) in een serie eendekroos vijvers op nitrificatie en denitrificatie (**Hoofdstuk 6**) is bestudeerd in twee achtereenvolgende fasen. Tijdens de eerste fase waren de 7 vijvers van de proefinstallatie geheel bedekt met eendekroos (*Spirodela polyrrhiza)*. Voor het begin van de tweede fase werd de bedekking met eendekroos verwijderd van de vijvers 1 en 3, om algengroei in deze 'open' vijvers mogelijk te maken. Het eendekroos vijver systeem werd gevoed met effluent van een full-scale UASB reactor die het huishoudelijke afvalwater van Ginebra-Colombia behandelde. Het systeem werd continu bedreven met een hydraulische verblijftijd (HRT) van 3 dagen per vijver en een totale HRT van 21 dagen. Totaal stikstof in het effluent was significant verschillend in de twee fasen, met 13.8± 2.9 mg TN l^{-1} (63 % verwijdering) and 3.7±1.5 mg TN l^{-1} (90%) voor de eerste en tweede fase, respectievelijk. Denitrificatie was het belangrijkste verwijderingsmechanisme tijdens beide fasen, en bedroeg 43.5 % en 76.2% van de stikstof in het influent, in de eerste en tweede fase, respectievelijk. Ammonia vervluchtiging en bezinking waren onbelangrijke processen voor de verwijdering van stikstof in beide fasen. Nitrificatie speelde een belangrijke rol bij de stikstofomzettingen in het eendekroos systeem en het werd bevorderd door de introductie van aërobe zones in de vijvers 1 en 3. Ook denitrificatie speelde een belangrijke rol voor de stikstof omzettingen en verwijdering. Ondanks de aanwezigheid van zuurstof in de waterkolom vond denitrificatie plaats, waarschijnlijk als gevolg van anaërobe microsites in de biofilms in het systeem.

Meer stikstofverwijdering zou bereikt kunnen worden in eendekroos systemen door de introductie van aërobe zones in de eerste vijvers van het systeem. Als het effluent niet hergebruikt kan worden voor irrigatie kan aan strenge effluent criteria voor stikstof voldaan worden door het gebruik van hybride eendekroos-algen vijvers met aanzienlijk kortere hydraulische verblijftijden in vergelijking tot systemen die geheel bedekt zijn met eendekroos.

Het effect van de diepte van de vijver op stikstof verwijdering in eendekroos vijvers is bestudeerd in **Hoofdstuk 7**. De proefinstallatie bestond uit twee series met zeven eendekroos vijvers in serie, met verschillende diepte en gevoed met effluent van een laboratorium schaal UASB reactor. Drie experimentele condities zijn bestudeerd: DSP1 met een diepte van 0.7 m en een HRT = 21 dagen, DSP2 met een diepte van 0.4 m en een HRT = 12 dagen, en DSP3 met een diepte van 0.4 m en een HRT = 21 dagen. Het systeem werd gemonsterd en bemeten voor pH, temperatuur en zuurstof profielen, organische stof verwijdering (BOD$_5$), stikstof omzettingen, biomassa productie en biomassa stikstof gehalte. De gemiddelde verwijdering van stikstof was 598 mg N m^{-2} d^{-1} voor DSP 1, 589 mg N m^{-2} d^{-1} voor DSP 2 en 482 mg N m^{-2} d^{-1} voor DSP 3. Ondanks de lagere verwijderingssnelheid van stikstof in DSP 3, werd in dat systeem een hogere verwijderingefficiëntie (44 %, 43 % en 62 % voor DSP 1, 2 and 3 respectievelijk) gemeten vanwege de lagere oppervlakte belasting voor dat systeem. Dit laat zien dat het gebruik van het verwijderingspercentage als een parameter voor vergelijking van systemen met voorzichtigheid gebruikt moet worden. De operationele parameters (in dit geval de diepte) moeten meegenomen worden in de vergelijking. Denitrificatie was het belangrijkste verwijderingsmechanisme voor de drie DSP systemen. Stikstof verwijdering door biomassa productie was het op één na belangrijkste mechanisme voor de drie experimenten. Diepte van de vijver lijkt geen effect the hebben op nitrificatie en denitrificatie. Nitrificatie lijkt te zijn gerelateerd aan de oppervlaktebelasting, terwijl denitrificatie gerelateerd was aan de beschikbaarheid van BOD. De vergelijking tussen twee vijver systemen met verschillende diepte, maar bedreven bij dezelfde hydraulische oppervlaktebelasting (DSP 1 en 2) liet dezelfde stikstof verwijdering zien in de ondiepere systemen als in de diepere systemen. Dit suggereert dat eendekroos vijver systemen ontworpen kunnen worden als ondiepe vijvers, zonder de oppervlaktebelasting en stikstofverwijdering te beïnvloeden. Stikstofverwijdering bleek meer bepaald te worden door de oppervlaktebelasting dan door de hydraulische verblijftijd.

Het meeste onderzoek is tot op heden gedaan op laboratorium schaal of in proef installaties. Voor het proces van technologieontwikkeling is het belangrijk om resultaten te bevestigen op praktijkschaal. In **Hoofdstuk 8** werd de efficiëntie van een eendekroos vijver op praktijkschaal vergeleken met een algen vijver op praktijkschaal voor de behandeling van effluent van een UASB reactor, onder verder gelijke condities wat betreft klimaat, configuratie, samenstelling van het afvalwater en belasting. Het experimentele systeem op praktijkschaal bestond uit twee continu bedreven vijvers. Eén werd bedreven als algen vijver en de andere als eendekroos vijver (*Spirodela polyrrhiza en Lemna minor*). Het volume van de eerste vijver was 225 m^3, een gemiddeld oppervlak van 322 m^2, L/W verhouding = 13.1 en een diepte van 0.7 m. Het afvalwater debiet was 19.7 m^3 d^{-1} voor elk systeem en de theoretische hydraulische verblijftijd was 11.5 dagen. De vijvers werden bemonsterd en bemeten voor de volgende parameters: organische stof (BOD$_5$), totaal zwevende stof (TSS), ammonium stikstof (NH$_4^+$-N), totaal Kjeldahl stikstof (TKN), nitriet (NO$_2$-N), nitraat (NO$_3$-N), totaal fosfor (TP) en faecale coliformen (FC). In de eendekroos vijver ontwikkelden zich andere condities in het aquatische milieu wat betreft pH, temperatuur en zuurstof dan in de algen vijver. De eendekroos vijver was efficiënter in het verwijderen van organische stof en de algen vijver was efficiënter in het verwijderen van stikstof. Denitrificatie bedroeg het grootste deel van de stikstofverwijdering in de algen en eendekroos vijvers. Het op één na belangrijkste mechanisme voor verwijdering van stikstof was ammoniak vervluchtiging voor de algen vijver en opname door de planten voor de eendekroos vijver. In het ontwerp van eendekroos vijver systemen moet speciale aandacht besteed worden aan de reactor configuratie en het stromingspatroon om een goed contact tussen de waterkolom en de eendekroos bedekking te bewerkstelligen en om hydraulische problemen te verminderen.

Praktische toepassingen

Zuivering van afvalwater kan veranderd worden in een aantrekkelijk, haalbaar en duurzaam alternatief door het combineren van anaërobe voorzuivering, eendekroos vijvers en algen vijvers. Het geïntegreerde systeem van UASB reactor, algen vijver en eendekroos vijver biedt de mogelijkheid om verschillende vervuilende stoffen uit afvalwater te verwijderen en om een deel van de waardevolle stoffen uit afvalwater terug te winnen in de vorm van biomassa of biogas. Het effluent kan geschikt zijn voor lozing of voor irrigatie, afhankelijk van de behaalde verwijderingsefficienties.

Het ontwerp en bedrijven van dit geïntegreerd systeem kan op twee manieren benaderd worden. Ten eerste kan men het terugwinnen van stikstof door eendekroos opname en door irrigatie. Ten tweede kan men stikstofverwijdering optimaliseren om ontvangend oppervlaktewater te beschermen.

Als het doel van de zuivering is stikstof terug te winnen, dan is de beste optie om de terugwinning van stikstof door het opnemen in eendekroos te stimuleren , samen met het reduceren van de stikstof effluent concentratie tot een geschikte niveau voor irrigatie. Een configuratie met een efficiënte anaërobe voorzuivering gevolgd door een serie vijvers die geheel bedekt zijn met eendekroos is dan aan te raden. Ammonium stikstof concentraties in het influent beneden 50 mg l^{-1} en pH beneden 8 is dan gewenst om remming van de groei van biomassa te voorkomen. De vergelijking tussen twee vijver systemen met verschillende diepte en dezelfde hydraulische oppervlaktebelasting liet zien dat dezelfde stikstofverwijdering in het ondiepe systeem bereikt kon worden als in het diepe systeem. Dit betekent dat eendekroos vijvers het beste ontworpen kunnen worden als ondiepe vijvers, zonder dat dit de stikstofverwijdering beïnvloedt. Ondiepe vijvers zijn makkelijker te maken, te bedrijven en te onderhouden en in het geval van vijvers met eendekroos kunnen ze beschouwd worden als een productie systeem voor gewassen.

Als het doel van de behandeling is om stikstof te verwijderen vanwege lozingseisen, moet een strategie aangenomen worden die gericht is op het bevorderen van denitrificatie. Meer stikstofverwijdering kan bereikt worden in eendekroos vijvers door de introductie van aërobe zones aan het begin van het systeem, wat een aanzienlijke vermindering van de hydraulische verblijftijd toelaat. Strenge eisen voor stikstof kunnen daarom gehaald worden met relatief korte hydraulische verblijftijden. De configuratie van het systeem, in het bijzonder de neerwaartse en opwaartse stroming lijkt een belangrijk positief effect te hebben op de denitrificatie.

Het opdelen van het zuiveringssysteem in compartimenten verbetert de efficiëntie. Bij het ontwerpen van vijver systemen moet extra aandacht besteed worden aan de reactor configuratie en het hydraulische stromingsprofiel, een goed contact tussen water en biomassa en het vermijden van kortsluitstromen en dode zones.

Voor het verdere proces van technologie ontwikkeling worden de volgende studies voorzien en aanbevolen voor verder onderzoek:

o   Toekomstige studies moeten gericht zijn op ondiepe vijvers met het oog op een verbeterde stikstof verwijdering door de terugwinning daarvan in de vorm van eendekroos biomassa. Ondiepe vijvers kunnen ook de kosten voor constructie van zuiveringssystemen verminderen.
o   Verschillend gebruik van behandeld effluent en geproduceerde biomassa moet onderzocht worden. In het geval van hergebruik van effluent voor irrigatie zal het voldoende zijn om de stikstof concentratie in het systeem te verminderen tot 15-25 mg l^{-1}. Het gebruik van de biomassa als aanvulling voor het voer van vissen en varkens is een mogelijkheid die al wetenschappelijk onderzocht is. Verdere studies zijn nodig om de haalbaarheid van deze optie te bepalen.
o   Voor veilige lozing van het effluent op open waterlichamen moet de stikstof concentratie in het effluent laag zijn. In dat geval kan de verwijdering van stikstof beïnvloed worden door de groei bepalende omstandigheden voor nitrificeerders en denitrificeerders, zoals de zuurstof concentratie of de beschikbaarheid van oppervlakte voor bacteriële aanhechting. Het is belangrijk om studies te doen om de beste combinatie van eendekroos en algen vijvers te vinden voor de verwijdering van stikstof. Het introduceren van schotten in de vijvers zal de beschikbaarheid van oppervlak voor bacteriële groei verbeteren en zal de hydraulische karakteristieken van het zuiveringssysteem ook verbeteren. Het beste aantal en de verdeling van de schotten moet onderzocht worden. Het recyclen van aëroob effluent naar de UASB reactor of naar de inlaat van de

eendekroos vijver kan een interessante optie zijn om de denitrificatie te stimuleren.

o De verwijdering van pathogenen zal beïnvloed worden door het gebruik van ondiepe vijvers, de aanwezigheid van aërobe zones en het opdelen van het zuiveringssysteem in compartimenten. Deze effecten moeten onderzocht worden om de verwijdering van pathogene organismen te optimaliseren.

Curriculum Vitae

Julia Rosa Caicedo Bejarano was born on 19th of February 1953 in the city of Cali, Colombia. In 1970, she graduated from high school, after which she enrolled in the Valle University in Cali. She obtained the B. Sc. in Sanitary Engineering in 1977. After her graduation she worked for the Regional Water Authority. In 1978 to date, she joined Valle University, where she is a Senior Lecturer in the School of Natural Resources and Environment of the Engineering Faculty.

In 1993, she was awarded a scholarship from a cooperative project UNESCO-IHE/DUT/UNIVALLE with financial support from SAIL, to study at UNESCO-IHE, in Delft. She got her post-graduate diploma in Environmental Science and Technology, with distinction in 1994 and her M. Sc in 1995. In 1997, she received the support from the Cooperative Project UNESCO-IHE/DUT/UNIVALLE to pursue her Sandwich Ph. D. study. During her Ph. D. study she lectured in the field of Environmental and Sanitary Engineering in under-graduated and M. Sc. programs at Valle University. She also supervised several M. Sc. projects in Valle University as well as at UNESCO-IHE-Delft.

Her address in Colombia is: Universidad del Valle. Apartado Aéreo 25360. Cali-Colombia. Telefax: 57-2-3312175.
e-mail: julcaice@univalle.edu.co